DIGITAL CITIZEN'S
— Guide to —
CYBERSECURITY

HOW TO BE SAFE AND EMPOWERED ONLINE

ANTOINETTE KING

We help busy professionals write and publish their stories
to distinguish themselves and their brands.

(407) 287-5700 | Winter Park, FL
info@BrightRay.com | www.BrightRay.com

ISBN:

No part of this publication may be reproduced, stored in a retrieval system, or transmitted in any form or by any means, electronic, mechanical, photocopying, recording, or otherwise, without permission of the author.
For information regarding permission, please contact the author.

Published in the United States of America.
BrightRay Publishing 2022

TABLE OF CONTENTS

DEDICATION, v
FOREWORD, vii
INTRODUCTION, xi

CHAPTER
ONE
What Is a Digital Citizen?, 1

CHAPTER
TWO
Sharing Is Caring, 11

CHAPTER
THREE
Digital DNA, 29

CHAPTER
FOUR
Cyberbullying, 47

CHAPTER
FIVE
Gone Phishing, 59

CHAPTER
SIX
Love At First Scam, 73

CHAPTER
SEVEN
Fake News or Newsworthy?, 91

CHAPTER
EIGHT
Striking the Digital Balance, 107

CONCLUSION, 119
ACKNOWLEDGMENTS, 127
NOTES, 131
GLOSSARY, 137
ABOUT THE AUTHOR, 143

DEDICATION

This book is dedicated to my parents. It is with the unconditional love of my mother and the work ethic taught to me by my dad that I have been able to accomplish so much in my life.

FOREWORD

My name is Jeffrey A. Slotnick, CPP, PSP; I am the principal of Setracon Incorporated, a 22.5-year-old, highly successful risk consultancy. I have had the honor to know and work with Antoinette King for several years. We were first introduced at an ASIS Leadership Conference. We have since developed an extraordinary friendship and have co-presented at several industry events.

Upon meeting with Antoinette and hearing her speak, I knew she was bound for greatness. Antoinette possesses several traits that immediately drew me to her story. She is brilliant and a self-made woman, having started her career as an installer of physical security devices—a career field that has been typically male-dominated. Years later, Antoinette learned the industry and represented a significant manufacturer of physical security devices, leading sales.

Over the past ten years, the physical security device industry has conjoined with the information technology industry to create and produce physical security apparatuses that now make up part of the cyber backbone and what is commonly referred to as the "Internet of Things" (IoT). What we used to call closed-circuit television is now network video. What started as being referred to as convergence now finds us converged.

Due to her knowledge of physical security technology and IoT, Antoinette was a leader in her company, researching and establishing programs to prevent cyber vulnerabilities. But even this was not enough for her. Antoinette attended university as a mature adult and obtained a master's in cyber security. Antoinette followed this major accomplishment as a driven risk-taker by opening a cyber security consulting firm titled Credo Cyber. Since that opening, Antoinette has received global recognition for her business actions and achievements.

I am a security management professional and very aware of cyber security. I have worked very hard to avoid the pitfalls of all the numerous exploits that have been used. I have stayed updated on the National Institute for Science and Technology (NIST) cyber security requirements. As a crime prevention specialist, I have even taught a course on practical cyber security for seniors.

In the past three years, my business, my mom, and my wife have all been victims of cybercrime. Scammers compromised my 89-year-old mother using a Microsoft technical support exploit, and she gave them access to her computer. That scam cost her $3000 wired offshore via Western Union and a week to change all her bank accounts and passwords and have her computer wiped and restored. They continued to harass her for almost a year after the event, trying to scare her into giving more money.

Earlier this year, my wife responded to a well-prepared phishing email, and her credit cards were compromised; the investigation is still ongoing.

In January, my business was attacked first with an attempt to spoof my company by scraping the web and creating an alternative website and new email accounts claiming to be my business partner and I. The action resulted in false invoices to several of our clients accompanied by payment instructions to an illegitimate bank account. It took over a week of working with local and federal law enforcement, working with the domain registry company to take down the fake domain, contacting the security department at the bank to shut down their bank account, and engaging a cyber security specialist to look for the compromise on all our on-site business systems.

Interestingly, we found no breach of our local systems. Tuesday of the following week, we suffered a denial-of-service attack that sent out 16,000 malicious emails to every person we ever contacted via MS Outlook. After two days with tier 3 tech support, we located the issue; hackers had found the logon and password for our Office 365 Enterprise Admin account on the dark web and used it to gather intelligence on us by changing our admin password and making changes to our Azure security settings.

Fortunately, none of our clients were compromised, and no money was lost. Reputationally, we suffered a setback that took us several weeks to resolve. The total time and funds we committed to resolving this event were very significant.

Post-event, I learned about why we should frequently change our passwords and what multi-factor authentication is, and I also learned about application passwords. Suppose I had read Antoinette's book *The Digital Citizen's Guide to Cybersecurity: How to Be Safe and Empowered Online*. In that case, I could have saved my company and myself a significant amount of time and money. Most importantly, I could have protected my company's reputation and prevented stress.

I strongly encourage you to read Antoinette's book; it is highly informative and simple to digest, and it provides you with practical steps you can take today to protect yourself, your business, and your assets.

Thank you,

Jeffrey A. Slotnick, CPP, PSP

President Setracon Enterprise Security Risk Management Services

INTRODUCTION

Life before the Internet was simple. I don't mean that in a "when I was your age, we walked three miles to and from school in the snow, uphill, both ways!" type of way. I mean that the present digital age introduces new, unprecedented issues and dangers almost every day. Thirty years ago, no one worried about digital addiction, fake news sites, or cyberbullying. None of those problems existed! At the same time, past generations never experienced the level of social interaction we now hold at our fingertips. With one tap of my finger, I can talk to someone from across the globe, learn about international conflict, and educate myself on any known culture.

Using our phones, laptops, tablets, and even our watches, we now connect with others via cyberspace every day, sometimes for hours on end. In other words, we are all digital citizens participating in a global digital community, whether we recognize it or not. In the physical world, we support our families, friends, and community members by visiting with them, being available to them during hard times, and attending special events in their honor. Similarly, in the cyber realm, digital citizens also support each other as we engage online. We need to use different tactics and resources to uplift and protect ourselves and those around us, also known as practicing good cyber hygiene.

Today, the idea of staying safe on the Internet isn't revolutionary or new for many people, but you want to know the crazy part? Most people on the planet can remember a time when the Internet, its opportunities, and its problems were still science fiction. Just flip through some 1980s sci-fi movies and you will understand what I am talking about.

Personally, I didn't grow up with a smartphone in my pocket or wireless earbuds in my ears. Despite that, I've always felt drawn to technology. Many people do not feel a passion for computer science until they enter college or begin their career, but I can honestly say that I felt interested in tech at nine years old.

I was raised in a classic Italian household. My mom stayed at home and cared for the family, while my dad owned and ran a body shop. Looking back on my childhood as an adult, I realized how my parents never upheld gender roles within our household, despite our traditional family environment. I have one sister and one brother, and we all completed the same tasks and chores. We dusted, vacuumed, and did laundry while also mowing the lawn and washing the cars. My dad took special care to teach each of us how to change a tire and replace car oil. It never dawned on me that boys should earn a living one way while girls should earn a living another way.

We were a single-income family, and my dad pulled 17+ hour shifts a day to support us. Putting it simply, video games were a luxury, not a priority. But we did own a household computer: the Atari 800. In lieu of purchasing games, my father would instead take trips to RadioShack to buy us booklets filled with video game code. My sister and I would connect our cassette player into the computer with a serial cable. Using the books our father bought us, we'd take turns typing in the code, letter by letter and number by number. Did we know what we were doing? Of course not, but we were good at following instructions (for the most part).

By the end, we learned a lesson about copying syntax: if you didn't enter in every value correctly, your game wouldn't work. The fun part was not knowing if you'd entered it right until you finished typing in several pages of code. My sister or I would submit the last few letters, sit back, and cross our fingers, hoping we produced an actual game. Nine times out of ten, we made a mistake. We'd try again, scanning through each line to find our error. It was tedious, but when it worked, man were we excited!

In fourth grade, my school introduced us to the Commodore 64, a large, clunky computer with a massive monitor surrounded in plastic. Though the computer itself was extremely bulky and took up a lot of space, the screen was a measly 12 inches wide, surrounded by a giant case. As part of an honors program in school, I was allowed to pair up with other students and use the computer for math games. With only six terminals available, we'd huddle together around the monitors, two kids to a computer, and answer math problems to advance through the game's levels. It was the mid-1980s; the Internet was not a "thing" yet. We didn't have Google or Internet access. The short, limited games were all we had, yet they were the most exciting part of our day.

Think of how quickly we evolved from floppy discs into smartphones. One time, my sister and I typed 300 lines of code to create Frogger. Imagine attempting to copy the syntax for a modern video game like Fortnite or Minecraft. Our current systems are far too complex for my childhood experiences to be replicated, and that change only happened over the span of two decades. Instead of crowding around a computer to solve algebra problems, kids now hold their own portable computers in their hands, and the capabilities extend way beyond playing games.

Technology has wildly accelerated since Baby Boomers or Generation Xers first began incorporating it into their daily lives. My father loves to tell me his experience with the Electronic Numerical Integrator and Computer (ENIAC) computer. He was fresh out of high school and eager to volunteer in the Vietnam War, but the U.S. military wouldn't enlist him because he had broken his ankle in a fight and was hobbling around in a cast. With his plans thrown out the window, my father decided to enroll in a computer course.

In the course, he learned to operate an ENIAC, also known as the very first programmable, digital computer. The ENIAC was designed to solve numerical problems. This meant that in order to operate it, my father and other participants had to use punch cards to plug numbers and letters into the machine, manually adding the zeros and ones we all take for granted today. My father had to use 50 cards just to type his name!

He would punch the cards and watch as they flew up the vacuum tubes. Then, the computer would spit out your name in DOS, an old computer operating system like Windows or Apple iOS but way less cool. Not only would the process take over an hour to complete, but if you messed up one of the cards, you would have to start over again. As you can tell, not much had changed from the time my dad started with computers in the late 1960s to when my sister and I were sitting at our computer desk in her room in the late 1980s.

In his lifetime, my father went from zero computers to the beginnings of computers to where we are now. Keep in mind: there's no beginner's course or how-to manual for all things technology. No, my father—and everyone else in his generation—adapted as new inventions, processes, and ways of thinking cemented themselves into our society. If you ask me, they don't get enough credit for that!

First came the ENIAC. Then, the Commodore and IBM's first personal computer came out, alongside the Atari and Apple computers. Every year, there was a new computer to hear and learn about. Each passing model offered brand new, unprecedented features until eventually, we arrived at our modern PCs and laptops. Even today, we experience similar growth in technology with new innovations coming out all the time. All the while, people are just expected to learn and operate the latest models seamlessly.

Oh, we also threw phones into the mix, too. My family owned a rotary phone until push-button phones became available. Using a rotary phone, it could take upwards of 45 seconds to a minute to dial a phone number, depending on how long the area code was and if you messed up a number while dialing. Whether I used a rotary or push-button phone, I'd still attempt to walk around the house while talking, causing the cord to tangle around and trip me. That's partially why the portable phone was so exciting, even if they were basically bricks with antennae.

They set the groundwork for later inventions and untethered communication.

Of course, you couldn't use them near microwaves. The earliest portable phones only had one frequency setting at 2.4 gigahertz. So, not only would you hear static if you turned on the microwave, but your neighbor could also tune their phone into yours and listen to your conversation, which mimicked our earlier problems with party lines. Over time, we progressed from the wall phone, to the portable home phone, to pagers, then to "car-phones" (the predecessor to the cell phone), to flip phones, and then finally to smart phones in the span of about 30 years. Even now, we're evolving past calling to texting. Heck, some people will choose Snapchat over both of those! Technological growth is never ending.

If you grew up immersed in technology, it's likely your brain operates differently than someone who did not. Your brain may be quick to figure out the next computer, phone, or app—after all, you were born into this way of life. On the other hand, if you once experienced a life void of the technology we now manage on a day-to-day basis, then you may feel like you've spent the last decade playing an endless game of catch-up.

Whichever group you belong to, we all have something in common: we are global, digital citizens who consume information and data in every aspect of our lives. In 2021, the average American spent 4.1 hours a day on mobile devices, a statistic no doubt impacted by Covid-19.[1] For the first time in the history of the human species, we are more connected than ever before, yet we report feeling more isolated than we have ever been.[2]

Despite the constant presence of the Internet in our daily lives, we somehow disregard the importance of critical thinking and cybersafety. We've all been thrown into this world of innovation without being given the tools to navigate it safely and intelligently. Without resources and knowledge, we haphazardly participate in online spaces, leaving us vulnerable to serious threats like cyberbullying, oversharing, scams, and misinformation.

By writing this book, I hope to empower you to reclaim control over how you engage online. With more than two decades of experience in the security industry, I guide digital citizens through easy and actionable cyber hygiene on a daily basis. To me, the Internet can be a gift. We all must learn to use it responsibly, both for ourselves and others.

After all, cybersecurity isn't simply your responsibility—it's *everyone's* responsibility.

CHAPTER ONE

WHAT IS A DIGITAL CITIZEN?

In high school, I took honors French. The best part of that class was the pen pal program where students sent letters back and forth with someone across the globe. I would bury myself in my textbook and learn French syntax, diction, and connotations to translate my thoughts into written letters, hoping to make them at least somewhat readable for my pen pal. Waiting in line at the post office just to hear back from my newfound French friend, I'd buy special stamps and envelopes for international mail so I could write my next letter as soon as I arrived home. For the first time in my life, I had the opportunity to connect with someone who lived on the other side of the ocean. It was a big deal!

The way we interact with others on a global scale has drastically evolved since my time in high school. I can't imagine a point in history when humans have been more in tune with one another. In our back pockets, we hold information about every culture on the planet, including cultures that don't even exist anymore. Long gone are the days of waiting for the newspaper to be delivered or the 5 o'clock news to come on TV to find out what is happening in the world. Today, many of us wake up in the morning, roll over in bed, and read the first headlines we see on Twitter to learn about what different people are experiencing across the globe. Thanks to satellites and GPS software, we can "travel" anywhere on the planet without ever leaving our homes. No matter how we use the Internet, our idea of community has undeniably changed.

CITIZENSHIP DEFINED

Before we can dive into what it means to be a digital citizen, we must first consider what it means to be a citizen at all. Think about every community you interact with in your daily life. The first group most people think of is their local community, which encompasses their immediate family, extended family, friends, neighbors, and other acquaintances who live close to them. This form of community is the one with the most physical, face-to-face contact. As a result, we tend to form deeper, more personal relationships with the people within our local communities.

Outside of our closest relationships, we may feel a sense of community with people who live in our state. If I told you to imagine someone from New York, you would probably envision someone with a distinct accent and assertive personality who takes pride in where they grew up. If a New Yorker moved to, say, California, they may feel a connection to other people they meet from New York simply due to their common birthplace. Then, as we expand broader, the people within our nation also act as a community. As an American, I have the same life experiences as a lot of other Americans, which allows me to more easily relate to them. If I lived in Canada, I wouldn't feel the same attachment to the American community.

Finally, we reach the broadest type of community: global communities. Though we will never meet every person on the globe, we still share a human connection simply because we inhabit the same planet. Even if we've never met someone in a different country, we can still feel real emotions for them during times of crisis or war. Humans are built to form relationships with one another. In a way, we are all citizens who work within communities to form bonds and engage with others.

Just as we define "citizen" in the traditional sense, we can expand the definition to fit our digital landscape. Instead of engaging with your neighbors, maybe you're part of an online forum about your city. Instead of joining a baseball league, maybe you frequently visit a Facebook fan group. For some of us, even our work communities have shifted to online spaces. In 2020, COVID-19 presented an environment for business owners to recognize how remote positions can save them money, increase productivity, and boost employee happiness. For better or worse, remote work has also challenged the traditional work environment. When working remotely, no one talks by the cooler or yells a joke across the office. Instead, employees forgo physical connection to communicate digitally, forming online bonds to act as a team and work together using video conferencing platforms and other means of digital communication.

Unlike a local, state, national, or global community, our digital community is borderless and constantly growing. Therefore, a **digital citizen** is anyone who actively participates in the use of technology and engages with others via cyberspace. In other words, anyone with Internet access is a digital citizen, whether they spend half of every day staring at their phone or if they only glance at Facebook once a week. Even people who claim to be "off the grid" still have data they're responsible for somewhere on the Internet.

Online community building presents so many unprecedented opportunities for global collaboration and understanding. We are lucky to live in a time riddled with limitless cultural exchange and knowledge even if we don't always think about it that way. And though the Internet provides us with so much to be grateful for, I'd be remiss if I didn't discuss the hard truths, too. Our limitless connections with other human beings also present their fair share of challenges. Checking the news every morning can exhaust us, leaving us emotionally drained about problems over which we have no control. The innate human need to be accepted and liked causes us to overshare private information and learn more than we'd care to know about some of the people in our lives. Even seemingly innocent interactions with online users can introduce us to harmful ideas, scams, or security threats.

On top of it all, our community leaders have shifted from local parents, teachers, and doctors to global influencers, celebrities, and YouTubers who share their thoughts on a wide range of topics. In fact, we expect these digital leaders to provide an opinion on almost every breaking news story or latest controversy. Makeup artists post political commentary on their Instagram stories, singers share links to charity organizations, and home chefs share their sympathies for global tragedies on their food blogs. These topics have nothing to do with their content, but as citizens part of a larger community, we look to them for guidance. We value their thoughts, no matter how little they know about the topic at hand. Just like how you trust your parents to provide you with solid advice on dating or changing a tire (even if they're not relationship therapists or mechanics), some digital citizens look to online community leaders for wisdom and insight.

Humans are social creatures. We evolved to form relationships and work together. The most impactful technologies in history always bring about great societal changes in how humans communicate and interact. Of course the Internet would be no different! Now, we just need to learn the skills necessary to keep ourselves safe, make sure our interactions actually benefit each other, and build more welcoming communities.

Baking the Digital Citizen Cake

As digital citizens, we share responsibility for other members of our communities. We have an obligation to create safer, more beneficial online spaces for the betterment of our current and future digital society. How does one go about doing that? In summary, think of it like the ingredients for baking a digital citizen cake. This is one recipe you will want to save!

Recipe For a Good Digital Citizen

Ingredients

- ¾ tablespoon **understanding of digital citizenship.**
- 1 ¼ cups **acknowledging that you're engaging with real human beings.**
- 4 cups **spreading positivity in your online communities.**
- 1/3 cup **avoiding desensitization.**
- 3 large pinches of **staying alert.**
- 1 stick of **fighting against misinformation.**
- 2 teaspoons **abstaining from knee-jerk reactions.**

Instructions

1. Prepare ingredients by **learning the definitions of digital citizenship and online communities**.
2. Combine **empathy for other humans** with **your online actions**. Mix to **uphold the same values online as you do in the physical world**.
3. Slowly sprinkle in **positivity** by **actively fighting against misinformation** and **avoiding desensitization**.
4. Pour mixture into your Internet practices. Add **staying alert** on top while ensuring you **abstain from any knee-jerk reactions**.
5. Bake for around the time it takes to read this book. Enjoy!

Being a safe and empathetic digital citizen means recognizing the people behind the screen. When engaging online, it's easy to see other people as profile pictures in a sea of content. By losing face-to-face contact, we can disconnect from one another, which can cause us to say or behave in ways we wouldn't in the physical world. We may even become desensitized to other people's feelings since we see so much negativity on a daily basis. However, by upholding the same integrity we have in the physical world, we create a safer and more forgiving online society. It's essential to recognize that when you engage online, you're not speaking into a void. You're communicating with a real person or real people, not a computer or social media platform. The way you'd behave at school, work, the grocery store, or any other public place is the same way you should behave online; you should carry the same behavior and ethics while engaging with others in a digital environment.

At the same time, one of the biggest mistakes you can make online is overly trusting someone and taking every piece of information at face value. It's a harsh truth to accept, but not everyone has your best interests at heart. Best case scenario, you learn this by clicking on a link for a news story only to end up watching the music video for Rick Astley's "Never Gonna Give You Up" (for all my readers who have never experienced this, the prank is called "Rickrolling," and it's actually pretty funny). Worst-case scenario, you are tricked into giving up personal information like your credit card numbers or sending someone money, only never to see it again (more on that later). Learning about all of the ways in which someone can access your information can seem intimidating, but it's imperative in our current digital climate.

Most importantly, remember to avoid knee-jerk reactions. Online spaces provide us with the unique opportunity to pause and think about what we're saying before we say it. How many times have you reread a risky text message over and over again before sending it? The same level of care should be taken with people you don't know. Pause, breathe, and consider the impact your words will have, not only on the person you're communicating with, but also on the other people potentially reading your conversation. In the same vein, you also have the opportunity to fact check before impulsively sharing a post or article. Anyone can create a blog nowadays and make any claim they want to. Performing your own research into topics can be an extra step, sure, but it prevents you from spreading lies or inaccuracies.

Overall, the best way to be a good digital citizen is to stay positive. I'm sure we can all agree that the Internet is way too overloaded with total downers. Between the news, YouTube comments, and Twitter arguments, every digital citizen could use some more positivity in their life. I don't mean you only have to share inspirational quotes or pictures of food to stay safe. Encouraging positivity means stopping to defend people being ostracized or bullied. It means informing others of scams, threats, or misinformation and tips on avoiding future dangers. It means taking the time to pause in your scrolling and searching to do what is right for someone else's benefit online. At the end of the day, that's the only way we can truly create a better, more unified digital community for the benefit of every digital citizen.

In short, it starts with you.

CHAPTER TWO

SHARING IS CARING

If you have ever been in a high school health class, you know about American psychologist Abraham Maslow and his "hierarchy of needs." In the mid 1900s, Maslow proposed a model that ranked human needs according to importance. Once someone fulfills one need, then they can move on to the next. In his famous triangle (See Figure 2.1), Maslow placed essential needs like food, water, and safety at the very bottom. Right after live-or-die requirements, he asserted humans need to feel love and belonging before they can ever feel accomplished or fulfilled in life. Basically, he said social connections are vital to happiness. If you're neuroscientist Matthew Lieberman, you may even say they are as important as other basic needs like food and warmth.[3]

Figure 2.1

The human brain is essentially hardwired to connect and share information because of this need for social interaction and acceptance. Scientists make new discoveries about the brain every day, making neuroscience a tricky field that constantly changes based on new insights. However, we do know some certain information concerning the ole noggin. For starters, the human brain features an abnormally large prefrontal cortex, which contains more neurons when compared to other primates.[4] Since our cortexes control how we behave around other people, express ourselves, and make decisions, many think our cortexes evolved to give humans a social advantage over other animals. By teaming up in groups, humans gained strength in numbers among a host of other mental and physical benefits.[5]

Our brains evolved to become dependent on social connections. As a result, the people within our families, communities, and cultures influence how we think, act, speak, and present ourselves. Loneliness impacts us not only emotionally, but also physically, even affecting how long we live. Likewise, when we have a productive conversation or interaction, our immune, endocrine, and cardiovascular systems all benefit.[6] We literally cannot thrive without communication and engagement.

In other words, humans are built to share. We share stories, recipes, personal likes and dislikes, facts about our pasts, and so much more to establish relationships vital to our mental and physical wellbeing. For this reason, it makes sense that the Internet, a powerful and far-reaching tool, would primarily serve as a vehicle for those connections. Using social media and other platforms, we can communicate with more people than we would ever be able to without it, enabling us to learn more, form beneficial relationships, and express ourselves in ways never seen before.

However, anyone who has so much as glanced at Twitter can tell you that while posting about yourself and talking to people online can be a great way to connect with others, it is also a double-edged sword. Despite what your aunt on Facebook or best friend on Instagram may say, oversharing online can be dangerous. Your online and real life personas can often feel like different people, but the truth is, they aren't. You may even browse the Internet believing you're posting anonymously to a website like Reddit or another forum, but in reality, you leave behind an identifiable footprint wherever you click. No matter if you share in the open or behind seemingly closed doors, the information you share online can very well lead to real life consequences, including identity theft, personal data monetization, and falling victim to scams.

We hear these buzzwords all the time on the news, but how do we actually go about preventing these offenses? Some would tell you to delete your social media accounts or look into going "off the grid." However, I'm not here to thwart your dreams of becoming the next big viral sensation. You don't have to delete your profiles or burn your cell phone. If participating online via social media, vlogging, blogging, etc. feels right for you, then you need to take the necessary steps to ensure your information is secure and avoid questionable apps and/or websites. I'm happy to say you can still be an influencer without posting all of your personal and private information to Instagram.

Ew, TMI: Personal vs. Private Information

It's easy to lose a sense of personal boundaries when posting online. Scrolling through any social media feed, you will see baby pictures, license plate numbers, and favorite restaurants. Some people will even post their current location for the world to see. The public sharing of any and all information seems commonplace. However, you wouldn't walk up to a stranger in a coffee shop and show them a picture of your dog or tell them your favorite color, so why does this become acceptable when we do it on social media? One of the most vital parts of being a good digital citizen is upholding the same ideals online as you would in person. This especially applies to the types of information you share.

The first step to becoming aware of your sharing habits is distinguishing the kinds of information you can share and understanding the consequences of oversharing. There are two types of information you can share on the Internet: personal and private. **Private information**, also known as Personally Identifiable Information (PII), is any information unique to you, such as your social security number or credit card information. Most of us know to avoid trusting random strangers online who ask for, say, details concerning our health or our home address. The thing is, many people have stopped asking. Instead, they find ways to sneak by you entirely and steal it out from under your nose, most often by using the second type of information, personal information.

Personal information is information not unique to you, such as the name of your school or workplace, your eye color, or your favorite sport. Sharing personal information could be as simple as talking about your favorite food or posting a picture of your pet. It seems harmless, and at one point or another, everyone has done it. Think about the one person you know who cannot have a single thought without posting it to Facebook first—we all have at least one of those people in our lives. They post what they ate for breakfast, how fast they ran in the afternoon, and what TV show they wrapped the day up with every single day. What's the harm in posting a workout video or two?

Unfortunately, sharing personal information can be just as harmful as sharing private information. Criminals, creeps, and all around dangerous people use the same Internet as you, but for them, it is the place where they search for their next victim. Online environments tend to encourage shocking behavior to garner likes, shares, and comments. Oversharing is the mechanism to achieve this type of attention, which in turn gives the bad guys everything they need to exploit that information, whether that be through data monetization, identity theft, or some other means.

Data Monetization

Have you ever wondered why websites like Twitter and Facebook allow users to create new accounts without charging them a penny? How can they be multi-billion dollar companies without requiring a subscription fee? The answer lies in a term called **data monetization**, or when a company sells your information to advertisers who then use it to create targeted ads. As I always say: if a service is free, *you* are the product. When we use websites and social media platforms, we're building millions of data points that companies then analyze and share with advertisers to make a boatload of cash.

Your data can end up in a company's hands one of two ways: legally or illegally. The first way becomes legal when you hit "Agree" on a website's terms and conditions contract, or as most of us know it, the giant block of text you skip every time you visit or sign up for a new website. In that contract, companies will have a section called their "privacy policy," which outlines what specific data they collect, how they use it, and who they share it with. Sometimes, companies will say they do not share data with any outside parties. In other cases, companies will state they share data with "affiliates," which is code for any other company that will pay them for their database.

The illegal method involves a mysterious part of the Internet called **the dark web**, a collection of hidden online content only accessible via a special web browser called Tor, short for The Onion Router. Think of Tor as an ordinary web browser like Google Chrome or Firefox except Tor allows its users to browse with complete anonymity. Whenever someone navigates to a web page within Tor, thousands of volunteers from around the world then route the web page request through proxy servers. This makes the IP address associated with the request untraceable. The anonymous nature of the dark web makes it a hotspot for illegal activity, especially when it comes to data theft. Though, it is important to note that some use the dark web for legal purposes, such as bypassing censorship or browsing the dark web's own Facebook-like social platform called BlackBook.

The first time many people hear about the dark web is when watching hacker movies and TV shows. Despite how pop culture portrays it, accessing the dark web is much more complex than simply visiting an off-limits website. On top of knowing how to access the dark web, you must also be accepted into the dark web community; otherwise, you will be blocked from participating in many of the forums almost immediately. Criminals use the dark web for a wide variety of tasks, including the illegal selling of information. Say a hacker breaches an organization's database; they can then steal the data and sell it on the dark web. The information is then used by other criminals to exploit their victims in various ways.

There are five main types of data criminals steal and sell on the dark web:
- Medical
- Financial
- Personal
- Government
- Business

According to Verizon's 2020 Data Breach Investigations Report, medical information is the most valuable form of PII.[7] An entire patient record could sell for as much as $1,000 depending on what information it contains.[8] In contrast, the price for financial information (such as credit card numbers) is lower due to the various methods credit card companies have for detecting fraud. Banks have essentially made the information unusable or, at the very least, not worth the effort of obtaining. Similarly, personal and government information are so easily attainable via social media and public records that most people do not need to buy them from the dark web, so their prices tend to be the lowest.

While you have no say in whether a hacker breaches a database or not, it is possible to opt out of sharing your data with companies. By opting out, you can limit how much of your information is out there to steal. However, the process can be extremely time consuming and involves contacting multiple data brokers to request they remove you from their databases. Some states, though, have laws that allow citizens to forbid companies from selling their data. For example, California passed the California Consumer Privacy Act (CCPA) in 2018. The law allows citizens to view exactly what information a company is collecting on them and opt out of any data mining and selling. New York also signed the Stop Hacks and Improve Electronic Data Security (SHIELD) Act into law in 2019 that requires companies to inform consumers in the event of a database breach and enhance data protection measures. Check your local laws to see what protections your state has in place. You never know; you may find a resource to better protect yourself against excessive data monetization.

Identity Theft

Identity theft is one of those crimes we may hear a lot about but never know what it's like until we or someone we know experiences it. By definition, **identity theft** is when someone wrongfully obtains and uses PII to commit fraud or other crimes. An identity thief could use stolen information to illegally apply for a credit card, file taxes, or receive medical services.

What we need to understand is cybercrimes are the same crimes that have existed for ages; criminals just use new tools to accomplish them. Identity theft is no different. One of the oldest identity theft tactics is collecting personal information on someone and using it to gain access to their bank account. Pre-Internet, criminals would swipe mail from other people's mailboxes and use the collected information to steal their identities. Now, there's no need to rummage through someone else's mail for PII when you can find it on the Internet with a couple of clicks.

When you set up a bank account, you are required to answer security questions to protect it. The questions are typically along the lines of "what was the model of your first car?" or "what is your mother's maiden name?" Most questions can be answered with a quick search through someone's social media, given they post frequently enough. All a criminal has to do is call the bank and determine which security questions are linked to the account. If they don't know the answers, they simply hang up, browse Facebook for a while, and then redial once they have the answers. Just like that, they will have access to not only your bank account (which is bad enough), but they will also be able to determine the social security number and address associated with the account. Wham bam, your identity just got stolen.

Oversharing on social media isn't the only way for someone to access your information. As we've discussed, criminals can also wait for someone else to infiltrate a database and, using the dark web, buy the PII they obtain. For example, a criminal may purchase someone else's Amazon username and password. Sure, the Amazon account would provide them with credit card information and their home address, but any criminal worth their salt will look at the bigger picture. Most people create accounts with the same username and password across all websites. Once you know them for one account, there is a good chance you know them for other accounts. With the Amazon username and password in their clutches, a criminal can then apply the same login information to a variety of standard websites like Netflix, Facebook, or any major banks. This tactic is called **brute-force attacking** and is extremely popular.

Have you ever received a message from a Facebook friend with odd wording and a suspicious link? Then, the next day, the same friend makes a post that says something along the lines of, "Don't open any messages from me. I was hacked!" This usually happens because a hacker brute-force attacked their way into the account. The most common question people ask when this happens is: "Why me? What could they possibly want with my account?" For the most part, these kinds of hackers are not interested in your specific information. They look at the bigger picture. They want your Facebook friends to trust your verified account and click the attached link, which is most likely

malware, or software that intentionally harms your device or compromises your account. Once your account is compromised, the hackers will have access to your friends' accounts as well. From there, they can keep sending the link and installing the malware on people's computers. Their end goal? They hope to create an army of bots to wield for much larger attacks and collect as much data as possible. Not only is your identity at risk of being stolen, but your infected device could also be used to harm others.

You never know how complicated having your identity stolen is until you're in that situation. There's more to it than calling the bank or post office and requesting they fix the problem. When every facet of your legal identity is altered, changing it back is a constant battle of trying to prove who you are to a never-ending list of people. Think of everything someone can do with access to your social security number and home address. They could open a bank account in your name, tank your credit score, and enter your information into all kinds of legal documents without your knowledge. Though the information you post online may exist in cyberspace, it can affect your life in a very real way filled with headaches, frustration, fear, and legal nonsense.

The best tactics to prevent online identity theft are often simple and actionable. For starters, every account you own should have a unique password, and I don't mean changing "Fluffy123" to "Fluffy124." Your passwords should be random in the truest sense, such as a password that looks something like AjV>j8Fz3d%?f8. To remember all of these different passwords, I recommend subscribing to a password vault service.

Password vaults are highly encrypted, online safes that hold all of your account information and generate strong passwords for you. Instead of remembering 500 different passwords for every account you own, you can instead remember one password for the vault. Then, you can also share passwords with people who you give access to the vault. Say you want to share your Netflix password with your family member. You can send it to him through a high security vault instead of simply texting it to him and trusting your phone to protect that information. With a password vault, you don't have to rely on Google Chrome to auto-fill your information every time you log onto a site, nor do you have to whip out your filled-to-the-brim notebook full of passwords. You become harder for hackers to manipulate, and because you're a target who takes way too much time to take advantage of, hackers will move on to the next person, potentially saving you from a serious identity dilemma.

In addition to using a strong, unique password, you should always use multi-factor authentication whenever possible. Multi-factor authentication, also referred to as MFA or 2-FA, is a login process where, in addition to a username and password, the user is required to provide a third factor for authentication. This factor could be something you have, something you know, or something you are. For example, Google and Apple have authenticator apps. When you log into an account using an authenticator app, you enter your credentials and are then prompted to enter a unique, timed code from the app as the third factor. This is something you *have*. Something you *know* could be a challenge question, which is an older method and not super secure. Finally, something you *are* could include the use of a biometric, such as fingerprint or facial recognition. These are often used with mobile device access. By implementing this third factor, you significantly decrease the risk of having your accounts compromised or hacked. Even if the bad actor has your username and password, without the additional factor, they will be unable to access your account, no matter how hard they try.

Stick to What—Or Rather, Who—You Know

The same safety lessons we learned as kids remain true when browsing the Internet. If a stranger approaches you, do not immediately trust or follow them. Stay away from dangerous areas; stick to the areas you're familiar with. Ask someone you trust for their insight if you feel uncomfortable or nervous, and seek help if you feel like you're in danger. The key to cybersecurity is upholding the same principles online as we do in real life. Just because a screen separates us from others doesn't mean we are free from danger.

As I'm sure you've picked up from reading the previous sections, some people on the Internet do not have your best interests at heart. I'm sorry to say this, but some users actually turn on their laptops or phones with the sole purpose of harming others, whether that be through means we've already discussed or in a more physical way. It is possible to manipulate personal information found on social media to find someone's home address, workplace, or school. By learning about someone else's hobbies, interests, and general personality, a person could even lure someone into a false sense of security and earn their trust, only to use it against them later. This is why if you haven't met a person face to face, you should never trust they are who they say they are.

My sole aim here isn't to scare you or make you paranoid someone is out to get you while you're watching cat videos on Youtube. My real intention is actually the exact opposite. By showing you that these dangers are real, I hope to provide you with the tools and knowledge you need to make smarter decisions while browsing the Internet. There is no need to look over your shoulder or stress over possibilities when you know you're practicing efficient cyber hygiene. Take the following tips to heart. Oh, and maybe think twice before posting that picture of your brand new car outside of your house. Just a thought.

Tips and Tricks For Protecting Your Information

1. Do not share any private information with individuals on the Internet, no matter who asks for it.
2. Be aware of the kinds of personal information you are sharing and with whom.
3. Recognize the dangers associated with oversharing information on social media.
4. Set up personal and public social media accounts.
    a. Set your personal account to "private" in the platform's settings to ensure only you and trusted users can view the content you post, share, and like.
    b. Use the public account to share general information that is safe for the public to see.

5. Enable **multi-factor authentication** when logging into your accounts.
    a. This will require the website to ask for two or more pieces of evidence that verify your identity before allowing access into the account, which prevents unauthorized users from entering the account.
6. Never share location data.
    a. If you have Snapchat, ensure you do not allow the app to track you with Snap Map, a feature that allows friends to see your location at all times.
7. Do not accept friend requests or messages from people you don't know.
    a. Delete these requests without even opening them.
        i. Some hackers utilize software that tells them if you've opened the message or not. If you do open it, they could view you as vulnerable, which would place an even greater target on your back.
8. Check your state's online privacy laws and learn how to opt out of data mining practices.

9. Download antivirus and anti-malware software to protect your devices against malicious attacks.
   a. I personally use McAfee and Bitdefender, but there are many options out there. As with any other service, do your research to see which software best meets your needs.
10. Consider subscribing to a virtual private network (VPN) service to protect your Internet connection and privacy online. VPNs mask your IP address so you can privately browse without being traced.
    a. Private Internet Access is a great VPN service, but many alternative companies exist, including ExpressVPN, NordVPN, and Surfshark.
11. Microsoft BitLocker is a Windows feature that encrypts all data on a hard drive to prevent unauthorized users from accessing it even if the hard drive is removed from the computer.
    a. This option is particularly great for anyone who works from home and has important business documents stored on their computer.
12. Subscribe to a password vault to document your account information, create strong passwords, and protect yourself against brute-force attacks.
    a. I personally recommend LastPass, but many services exist. Like with any software or extension you download, ensure you do your research on its credibility.

13. When creating security questions, do not answer the questions truthfully. Instead, create unique answers to the questions in order to prevent hackers from using social media to gain access. Remember to keep your answers somewhere safe like a password vault!

    a. For example, suppose the security question for my bank account asks, "What was the model of your first car?" Instead of writing "Toyota Camry," the truth, I may lie and write "Ford Mustang."

        i. Then, if my phone's account requires me to answer a security question in the future, I would not choose the same question again. I may pick "What was your first pet's name?" instead and make up another answer. Rinse and repeat for every security question you encounter.

CHAPTER THREE

DIGITAL DNA

There's a common phrase parents teach their children when they first begin to use the Internet: what goes online, stays online. The saying may sound like an exaggeration at first. I mean, when you click "delete" on a file or post, you never see it again, so it must not exist, right?

In actuality, much of what you post online indefinitely stays online in some form or another. As soon as you click the post button, the content no longer belongs to you, and you cannot control where it ends up or who uses it. This is, in part, beneficial to the constantly updating collection of online knowledge and resources. If nothing is deleted, then no information can be lost to the throes of time. However, this also means those TikTok dances you uploaded last year, cringed at, and hurriedly deleted could still exist. The embarrassing baby photos your mom posted could remain, as could all of the status updates you posted when you were bored. Think of everything you've ever posted online. Depending on who had access to it at the time you posted, your content may still remain somewhere.

Web Browsers vs. Search Engines

Though some people tend to use the name for web browsers and search engines interchangeably, the two are completely different tools. For example, you can open Google Chrome, a web browser, to then search on Google, a search engine. However, you could also open Firefox, a web browser, to search on Bing, a search engine.

Web browser: A software for accessing the Internet, such as Google Chrome, Firefox, or Microsoft Edge. A web browser allows users to open and navigate through the World Wide Web.

Search engine: A software designed to search the Internet for results that correspond to the keywords typed into the search bar, such as Google or Bing. A search engine allows users to find websites and information on the World Wide Web.

Your **digital DNA**, also referred to as your digital footprint, is the trail of information you leave behind when browsing the Internet. Even if you think you're navigating through cyberspace with complete anonymity, you definitely are not. As soon as you open your web browser, you add to your own personal treasure trove of data points based on all of the content you've posted on social media, the websites you have visited, and the ads you have interacted with. Unless you plan on living completely off the grid and abandoning all use of technology, digital DNA is permanent and cannot be removed, just like the DNA that makes up our physical bodies. The difference is that instead of being made of nucleotides and genetic information, digital DNA is built from all of the information gathered about you when you are engaging online.

Physical DNA    Digital DNA

When posting on forums, hitting "enter" on search engines, creating a blog, watching videos on social media, or interacting with any other online system, you generate millions of data points that say something about who you are and what you like. In the biz, we call these data points **metadata**, or the information that is stored about you, including your IP address, physical location, computer hardware, router type, personal information, and other identifying characteristics.

Say you type "Where can I buy red shoes?" into Google's search bar. The search will then be added to your web browser's perception of your interests, and as a result, you may see links to buy red shoes everywhere you click. In addition, the way you phrased the question and the words you chose all provide your web browser with an understanding of your mood, which it will then add to its understanding of how you feel about red shoes. Imagine you type in "High-end restaurants near me" right after your search for shoes. Based on this, your browser knows you feel like going out for a fancy dinner, so it may suggest formal high heels or loafers for the occasion. Likewise, if you typed in "Hiking trails near me," your browser may instead recommend athletic sneakers or boots. From the moment you begin engaging online, whether you interact with ads, websites, videos, or any other content, you leave behind metadata that shows your online searching and behavioral habits. In combination with metadata, the Internet also paints a picture of who you are through the use of cookies.

Cookies are small text files that are dropped into your browser when you engage on a website. These text files hold information about you to allow websites to remember you. Every time a cookie is created, data is stored in it. Then, that data is labeled with an ID unique to you and your computer. Have you ever wondered how a website remembers what you added to your online shopping cart, or how clicking a "Remember Me" box on a login page stores your username and password? The answer is cookies, convenient tools for website creators to track and save information about how users engage on websites.

Despite its delicious name, cookies can lead to a nefarious invasion of privacy. The collection of cookies becomes concerning when the garnered information is used for purposes other than website and ad personalization. As an example, throughout the 2010s, a British consulting firm called Cambridge Analytica illegally harvested data from around 87 million Facebook profiles without the owners' consent.[9] In 2016, the company then used the collected data for political advertising across Facebook, aiding in the spread of misinformation. As a result, Facebook paid a $5 billion fine to the Federal Trade Commission for privacy violations, Cambridge Analytica filed for bankruptcy, and the scandal jumpstarted a serious conversation about online privacy regulations and how digital DNA can impact large-scale, real world issues.

Pull out your phone or computer and visit any website. I bet there's a box somewhere reading, "Accept All Cookies." This box was originally driven by European Union legislation called the General Data Protection Regulation (GDPR), laws created to protect EU citizens from data theft and give them some control over their data. A key element of GDPR gives EU citizens the right to opt out of the collection of cookies. Though the law applies to the EU, all companies with global websites must now feature a clear way for users to forgo sharing their cookies or potentially face a strict fine, which is how the "Accept All Cookies" button came to be. While the button is intended to give users more control over the data that is collected about them, the buttons are so common that people blindly accept the cookies, not understanding what a cookie is or the choice they're making. Companies are actually *more* empowered to use your data now since your approval upon clicking the button legally justifies their actions.

No matter how much control these companies have, though, there are ways to protect your data and information while browsing the Internet. Personally, I have a habit of only opening my web browser in a private window. Google Chrome calls this private window "Incognito Mode," while Firefox calls it "Private Browsing." By opening a private window, you prevent the web browser from collecting your metadata, cookies, and other information. As much as you try to browse privately via a standard browser, your information will still be collected if you do not open a private window. In addition, you can also use a service like DuckDuckGo, which is a search engine with a specific focus on protecting its users' privacy. With these tools, you do not have to blindly browse the Internet and give up your cookies to every website you visit. It is possible to protect your information and still enjoy the wonders of the Internet.

Social Media Platforms' Terms and Conditions

How companies collect and use your data through their websites is detailed in the site's terms and conditions or privacy policy, but how many of us stop to read through every one of those we encounter? Chances are, we don't even have enough time to read a single one. Once, in a course on data privacy, I thoroughly researched Facebook's privacy policy, which looks quite benign on the surface—that is, until you start clicking on all of the hyperlinks. Suddenly, the simple two-page document leads you down a complex rabbit hole of legal jargon and hidden policies. Upon clicking one hyperlink on the first page, I was led to 14 other pages of information. If I took the time to click on every hyperlink afforded to me, I would probably visit hundreds of pages full of information, a lot of important details buried by the previous pages.

One such detail is this: Facebook reserves the right to maintain any and all information you shared on their platform, even if you delete the post or your entire account. In the event that your account is deleted, then the content you shared on someone else's page or via Facebook Messenger will remain. Therefore, certain content will be deleted while other content may still exist. Facebook is not the only social media platform that stores your data. Apps like Instagram and Snapchat also include some convoluted or vague policies in their agreement forms. As we've discussed, digital DNA is forever. The digital world may seem inconsequential to your real life, but the material you post online can always come back to affect not only you, but also your family and friends.

Surprising Policies Found in the Most Popular Social Media Platforms' Terms and Conditions Agreements

- *Facebook, Facebook Messenger, Instagram, and All Meta Platforms*[10]
  - When deleting a Meta account, you have three options:
    - » Delete your account with the option to undelete it within a set timeframe

- » Deactivate your account so it appears deleted but isn't
- » Permanently delete your account
- ◆ If you delete your account, you have the option to restore the account within 30 days. After 30 days, you will not be able to retrieve your account. However, the platform will then take up to 90 days to delete all of the information you posted. From there, copies of that information may remain in Meta's backup storage for another 90 days.
  - » This means it could take three quarters of a year to delete your information off of Facebook or Instagram entirely.
- ◆ Even if you choose to permanently delete your account, certain information remains on the platform, including:
  - » Messages you sent to friends
  - » Posts made to groups
  - » Posts shared to your friends' pages[11]

- Twitter[12]
  - ◆ If you decide to delete your Twitter account, the platform offers a 30-day deactivation window that allows you to restore your account within that given timeframe.
  - ◆ Deleting your Twitter account does not delete your information from search engines like Google or Bing since Twitter does not control those sites.

- » Though, you can send requests to search engines to remove the information.
- ◆ When you deactivate your account, mentions of your username in other user's Tweets will still exist. However, the mention will no longer link to your profile as your account will no longer be available.

- *Snapchat*[13]
  - ◆ Snapchat retains your information 30 days after deactivating your account. After 30 days, your account will be permanently deleted.
  - ◆ Snapchat stores your location data for various lengths of time depending on the services you use. If there is location information associated with a Snap, the app will store that information for as long as you keep the Snap.
  - ◆ You can download a copy of all of the data Snapchat stores about you using their "Download My Data" tool found in their privacy policy.

Sharenting

When my son was about seven or eight years old, he idolized Albert Einstein. Encouraging his love for science and physics, my husband and I bought him a plasma ball, one of those spheres you can touch where the lights follow your finger. My son soon discovered that if he touched the plasma ball to his Darth Vader alarm clock, the alarm would ring. When Halloween rolled around, my son, of course, dressed as Albert Einstein and prepared a whole presentation to announce his scientific discovery. I recorded the whole show, watching through my phone screen as he touched the ball to the clock and excitedly announced the result. Not thinking anything of it, I posted the video to YouTube so I could easily share the experience with my close family and friends, who also "oh" and "ah"ed over his adorable antics.

Flash forward to my son's middle school days. One of his friends discovered the video online and, thinking it was the funniest video ever, shared it with a bunch of other kids. In a few short weeks, the video spread throughout the school until my son couldn't go a day without hearing about it. The video's legacy even carried on until he entered high school. One day, he sat down to watch the typical morning announcements, only to see his Einstein wig and lab coat video being played to every student and teacher in the building. Though the people who placed the video on the morning announcements had no ill intentions, my son still felt mortified.

I was devastated for him. I immediately deleted the video. However, unbeknownst to me, one of his friends had downloaded it and continued to play it in front of other people. At the time of recording the video, I had no idea it had the potential to spread so far or cause so much humiliation. I made a video out of love, yet it somehow escalated into hurting my son. Watching him parade around as Einstein made me proud and excited, but I couldn't imagine how he would feel seeing the video later as a teenager. I created a video that documented his vulnerability without his permission. I succumbed to the phenomenon nicknamed "sharenting."

First coined in 2013, **sharenting** ("share" + "parenting") is a term used to describe when a parent publishes content surrounding their children on social media or other online platforms. Thanks to sharenting, an entire generation of people will have a digital footprint before they are even born. Without the child's consent, parents post their personal and, in some cases, private information for the whole world to find and exploit. A typical Facebook post about a newborn will often show pictures from before birth with ultrasound and gender reveal party photos. Then, after the child is born, a post is made that contains the infant's full name, birth weight, birth date, time of birth, and picture. Then, as they grow up, their parents will post every notable milestone, which could include their school names and graduations, their first car, the city they grew up in, their pets' names, their favorite foods, etc. Now think back to the most popular security questions and what they ask. How many security questions do you think a criminal could answer about the child if given access to their parents' social media accounts? I'd bet almost all of them.

Pre-Internet, proud mothers and fathers carried around wallets with multiple photo holders. If someone asked about their children, they could pull out their wallet, unravel the multiple photos, and show them to people in passing. If a child celebrated a birthday, had a wedding, or graduated high school, you had better believe all of their relatives would receive a wallet-sized photo of the event. It's always been human nature to share and make connections. Now, we just do it with our phones instead of our wallets. The only downside is that physical wallets are much harder to steal and misuse than online information.

Wallets also did not subject children to a large, global audience like social media does. Being on full display for hundreds, if not thousands or even millions, of people from a young age can have lasting effects on someone. According to a 2021 study, 93% of moms use a social networking site or service. While most of the surveyed moms primarily use Facebook, TikTok showed the biggest increase in usage among mothers, rising from 8% in 2019 to 26% in 2021.[14] With such a large social media presence, it's no wonder that average parents post nearly 1,500 pictures of their children online by the time they're five years old.[15] Like with all information shared online, it's not easy to control where those photos end up. Even if you delete them, someone could have them saved in their files, taken a screenshot, or shared the photos around the Internet, which would then further their reach. Plus, the metadata still exists, and depending on the social media platform, the website may still have rights to them. Once it's online, it's potentially there forever.

One of the most notable instances of mistaken sharenting is the story of David Devore Jr., or as many may know him, David After Dentist. When David's father posted a video of him after having his wisdom teeth removed, he never expected it to go viral in the way that it did. As of the writing of this book, David After Dentist has over 140 million views on YouTube. At the time the video was popular, David Jr. became so famous that he traveled around the world for talk shows and celebrity events, and the money his family earned eventually allowed him to pay for college. However, his family also experienced the dark side of the Internet. David Jr.'s father received multiple threats and accusations of child abuse, and the video continues to follow him and impact how he lives his adult life. Everywhere he goes, people see David After Dentist first, not just David.[16]

Sharenting is an easy trap to fall into. I, David Sr., and so many other parents have participated in the act and regretted it in the long run. We see social media as the modern day scrapbook. It's normal to document the lives of our children. We want to look back when we're older and see them take their first steps again or walk down the aisle. However, when clicking "post" on a picture of our child, we don't think about cleaning up the residual mess that accompanies our actions. Not only does "sharenting" expose our children's vulnerabilities and subject them to data theft, but what we share about them could also affect their future opportunities.

The Real Consequences of Digital DNA

I once participated in hiring new employees for a security firm I worked for. We dug through resumes and interviewed candidates, hoping to find someone who fit the job description and our work culture, the usual. My team was excited about one candidate in particular who perfectly matched the criteria. She had excellent references and a ton of experience, and her resume demonstrated that she had the skill set we were looking for. We couldn't wait to schedule her for an interview and hopefully pull her onto the team. Even a decade ago when this story took place, we had the foresight to scan through her social media first. There, we saw all kinds of pictures of her partying with her friends and acting in unprofessional ways we did not want associated with our company. We shredded her resume that day.

What you post online can affect you in real ways. One TikTok dance, profane Tweet, or opinionated blog post can impact so many future opportunities. Today, the first action a potential employer takes is searching your name on Google and all social media platforms. A new romantic partner may browse through your social media to better understand your interests and humor only to find a red flag. College rejection letters could come flying in when a university looks you up and finds PR nightmare after PR nightmare. Your online reputation can potentially influence almost every aspect of your life. Carefully choosing what you share on the Internet is an essential consideration. Imagine what importance it will hold years from now when online life becomes even more ingrained in our societies. Long story short, the trail you leave behind matters, especially when someone could follow that trail to put you or your future in danger via data monetization, identity theft, or even cyberbullying.

Tips and Tricks for Monitoring Your Digital Footprint

- Be intentional with the content you post online. Maintain the mindset that everything you interact with, share, and post will remain on the Internet forever.
- Understand how metadata and cookies contribute to your digital DNA.
- When conducting searches, try Google Chrome's "Incognito Mode" or Firefox's "Private Browsing" mode. These features allow you to browse privately without the collection of browsing history, website cookies, and third-party cookies.
  - You may also try installing DuckDuckGo, a popular browser extension that emphasizes user privacy. This search engine does not collect or share personal information.
- When you hit "Accept" on a terms and conditions agreement, make sure you know what you're agreeing to.
- Avoid sharenting, no matter how cute your baby looks or how many likes your relatives will leave on the post. Recognize that anything you post about your child could remain on the Internet forever and may even impact them down the line.
  - Some parents do not include their children on social media until they reach a certain age. For example, a mother may announce that she's had a baby but will not state the baby's name nor any personal information.

> » Ask yourself if this approach would help your own child avoid feeling exploited or exposed, too. Consider the future emotions of your children before including them in any online content.
- If you plan to post personal information on your social media, lock down your accounts as much as you can. Implement the strongest security the platform allows.
  - As of 2022, Facebook now only allows you to set who can see your profile to "Friends of Friends," not "Friends Only." This means anyone within one degree of separation (meaning, a friend of a friend) who clicks on your profile will be able to see your posts. Consider this possibility as you participate on the website.
- Disable the location settings in your phone's settings to prevent apps from tracking you. As a fair warning, the process of ensuring you have unticked all of the boxes can be a tedious one.
  - Even if you are not actively looking at an app, it may still be tracking your location if it is running in the background. Navigate to your phone's settings to disable its permissions, and as a precaution, always fully close apps that have the potential to do this.

- Before posting content online, ask yourself:
    - Will this come back to haunt me in my social life?
    - Could this impact any academic, professional, or personal opportunities now or in the future?
    - Does this hold any personal or private information I do not want to be readily available?
    - Is this something I would be okay with posting on a billboard in Times Square?
    - Am I posting information about someone else, either a friend, acquaintance, family member, or child, without their consent?

CHAPTER FOUR

CYBERBULLYING

One afternoon, I received a text from a professional colleague of mine who was struggling with a dilemma. The colleague, who we'll call "Henry" for the sake of clarity and anonymity, is a highly respected member of the cybersecurity community with more than 20 years of experience in the field. There have been times when I've reached out to him for advice or counsel only to receive some much needed wisdom. All of this being said, you would have never guessed that Henry received his formal education in the culinary arts, not in information technology or security systems.

This isn't an unusual event to see in everyday life. After all, most of us know that when you earn a degree in one area, there is a strong possibility you will end up doing something completely different with your life in the long run. So what if Henry could flip an omelet better than the rest of the cybersecurity community? He still knew his stuff from decades of hands-on experience.

Despite Henry's strong presence in respected cybersecurity circles, he was experiencing an onslaught of backlash from an individual online. Henry had posted an educational tidbit of information to his LinkedIn, detailing an aspect of cybersecurity he thought his connections would benefit from. Over the course of several weeks, he began receiving several private messages a day from an online user who claimed Henry did not have the right to speak on the topic of cybersecurity due his educational background. In addition to the messages, the user then began commenting on his posts, attempting to "expose" his lack of formal education to his followers and discredit him as an authority. No matter what Henry tried to do, the person relentlessly commented and messaged day after day.

When Henry reached out to me, he provided countless screenshots of the harassment, feeling distraught and not knowing what steps to take next. After scanning over the messages and comments, I provided Henry with one piece of advice: block and report this person, and put the situation to bed. Then, once he blocked the individual and deleted most of his comments, we gathered our group of security professionals and encouraged them to also block and report the offender. As industry leaders, we all came together and refused to tolerate such negativity in our community. If the perpetrator left any hateful comments, we ignored and buried them with positive, helpful statements instead. We avoided feeding into his argumentative nature, which ultimately granted us enough control over the situation to put a stop to it. My group and I took the oxygen out of the room instead of fanning the flames, meaning we purposefully did not enable the behavior and chose to replace it with positivity—which, I'm sure we can all agree, the Internet can always use a bit more of.

What my friend Henry experienced is called **cyberbullying**, a situation in which one person feels threatened, targeted, or harassed by another person when engaging in online spaces. When many of us picture cyberbullying, we tend to imagine teenagers or young adults. In reality, cyberbullying can happen to, or be perpetuated by, any digital citizen via any online medium, as we saw in Henry's case. Sitting down at a computer is no longer required—cyberbullying can happen through texting, apps, and video games. Someone could even leave a nasty Twitter reply or send a harmful text message through their watch! The stereotypical idea of cyberbullying (i.e. the teen who receives hurtful messages in an online forum but then can turn their computer off to separate themselves from the comments) is no more. With how essential technology is to our lives, victims of cyberbullying can no longer "just turn their phones off and ignore the haters." Cyberbullying now has the potential to influence every aspect of the victim's life.

The Effects of Cyberbullying

Unlike when we do something embarrassing in front of a group of friends or say the wrong thing during a work meeting, posting something we regret online can have large-scale consequences. The traditional idea of bullying usually involves the bullied and the bully, one person or a group of people who emotionally and/or physically attack a single target. However, cyberbullying involves many more parties. Online, a significant amount of people can jump into the conversation and add fuel to the fire, either by enabling the negativity or drawing attention to the target.

In short, cyberbullying differs from physical bullying in that it's much easier for groups of people to contribute online than it is in the real world. For example, if someone posts a video of themselves cooking a new recipe, a cyberbully may choose to comment on the poster's appearance rather than the food they're cooking. One comment can shift the entire conversation from the recipe to the original poster's body type, voice, or any number of personal characteristics the poster never meant to be discussed.

Suddenly, the video's comments, which could be a few or several thousand depending on if the video goes viral, only feature people talking about the poster. Reading several comments about themself in a row, the person may begin to feel insecure, which can result in them taking down the video, not wanting to post anything else, and losing out on an experience that they thought would be fun and rewarding. The video could then follow them into their real life, impacting how they view themselves and even contributing to their conversations with others. In the worst-case scenario, the people who commented on the video could begin bullying them in other online or physical spaces, which could lead to a decrease in mental health and impact their overall well-being.

When the term "cyberbullying" first entered everyday vernacular, most people didn't understand how it could possibly be a serious issue. Online harassment was an unprecedented phenomenon. As a result, many did not believe that cyberbullying could be as harmful as physical bullying. However, cyberbullying can be just as, if not more so, detrimental to a victim's mental health and can potentially lead to depression and anxiety, low self-esteem, and in many cases, self-harm and suicide. While a victim in a traditional bullying scenario eventually leaves where they're being bullied, whether that be school, work, or another area, and returns to a safe space, a cyberbullying victim carries their bully in their pocket with them wherever they go.

On top of the inescapability of cyberbullying, the posting of **hate speech**, or any public statement that encourages discrimination against a group of people based on their identity, can impact how online users see themselves and others without people even interacting with each other. Hate speech could be as blatant as saying hurtful comments to someone based on their race, ethnicity, sexuality, gender, or other characteristic, or as underhanded as supporting harmful generalizations about a group. Either way, if a person reads hate speech and feels targeted, they will feel discriminated against and victimized no matter if the original poster and victim have ever met or talked to one another.

It is this level of disconnect that enables users to perpetuate cyberbullying. When online, it's easy to become desensitized to negativity and complacent in its spread. Scouring the Internet, we often have to filter through and ignore terrible posts; after all, we're bombarded with them every day of our lives! If we stopped to read every negative thing we saw on the Internet, we'd probably open Twitter way less often. Scrolling through, avatars cease being real people and become little pictures on a screen to save ourselves from social exhaustion. People with inclinations towards bullying in the real world suddenly become more enabled, not feeling as if their words have real impact and not feeling socially prohibited from spreading rude comments, lies, or hate speech. When more online users see the cyberbully's actions and recognize that they have faced no consequences, the mob mentality kicks in, and people either jump in to defend the victim or contribute to the abuse.

The first step to rejecting desensitization starts within yourself. Every time you put yourself out there on a public platform, whether that be through a selfie or an opinion, there's always potential for someone to respond negatively. You must train yourself and set an example for others to respond appropriately, utilize effective communication skills, and have the courage to defend your personhood and beliefs. The fight against cyberbullying may take confidence and work, but fortunately, there are actions you can take to put a stop to it within your circles.

What Can You Do?

When Covid-19 caused everyone to close their doors and open their laptops, the entire world's daily interactions became almost purely digital. In addition to the many global dilemmas the pandemic caused, it also greatly contributed to the rise of cyberbullying rates. According to a study conducted by the Cyberbullying Research Center in 2021, more K-12 students reported having experienced cyberbullying than in previous years as in-person bullying rates significantly dropped.[17] Hateful interactions between children and teenagers in online chatrooms increased by 70% with a 40% increase in negativity across popular gaming platforms. Additionally, hate speech drastically increased in 2020 with a 900% increase in hate speech against China and Chinese individuals.[18] In short, Covid-19 permanently changed how we view technology on a global scale, which means this increase in negativity isn't going anywhere soon. So, with cyberbullying on the rise, how can we digital citizens add positivity to our online environments and stand up for others? I'm so glad you asked!

1. **Find the courage to oppose the mob mentality and speak up to support victims.**

   First and foremost, the fight against cyberbullying starts with being an upstander, not a bystander. As the names suggest, a **bystander** stands by and takes no action when seeing online bullying, while an **upstander** stands up for others and takes action to defend and support victims. It takes immense courage to go against the grain and stick up for someone else. Sometimes, it may seem almost impossible to be the only one who voices their opinion and says, "Hey, this isn't right." Not jumping on the bandwagon and aiding in mob mentality takes personal critical thinking skills; you have to be able to pause before acting to assess your own morals and determine the correct response. Thinking about both sides of the story and extending empathy is something you cannot do with a split-second decision, yet doing so is necessary for creating a positive online environment to which everyone feels safe contributing.

   If you see someone being targeted or attacked online, extend a helping hand. Construct a well-written response to the aggressor detailing how they could have better approached the situation, and stick up for the victim. Avoid engaging in any further conversation with the aggressor even if getting the last word seems enticing. Instead, get the last word by reaching out to the victim and expressing your support, showing them that they're not alone. At the end of the day, if your moral compass aligns with uplifting others, then you have to plant the flag of being a positive person and stick to it. A good digital citizen knows that sentiment extends to online communities, too.

2. **Only participate in platforms that foster positivity.**

Thanks to modern day algorithms, your social media feeds usually reflect what you're most interested in and who you want to hear from. As we've discussed, this can be tricky when dealing with data monetization and information sharing. There is a positive side to it, though: you control what you see when you browse online, meaning you can choose to surround yourself with positivity and productivity rather than hate and criticism. Like, share, and engage with content that makes you feel empowered and excited rather than inferior or anxious. Every post you like is a step to creating an online space that benefits you and your followers.

Consider scrolling through the list of people you follow. How many are constantly commenting on other people, making hurtful jokes, or even sharing ideas that are thinly veiled hate speech? How many promote content that uplifts their audience, such as body positivity imagery, mental health advice, or congratulations on accomplishments? Depending on how you answer these questions, it may be time for a following purge!

3. **Be the kind of leader in your group of friends or colleagues who makes it clear that bullying and harmful behavior will not be tolerated by you.**

   When you first started reading this book, I warned you that being a good digital citizen requires courage—courage to take responsibility, be there for the people you care about, and foster a healthy online space that makes the Internet a better place. Being a leader who sparks positive change means caring about other people's feelings, celebrating everyone's differences, and ensuring everyone knows you uphold these principles. Then, if your friend or professional group does encounter conflict, you must learn the best ways to handle it without turning into or encouraging the aggressor. The key is to channel negative feelings into helpful dialogue that allows everyone to see each side of the story and arrive at a productive solution. Have the courage to speak up about harmful behavior even when no one else does. It sets a precedent for everyone else and makes it easier to move forward.

4. **Know how and when to apologize for your own words or actions.**

   The Internet has a special way of painting every scenario in black and white. Are there some truly despicable characters out there on the World Wide Web? Sure there are. However, most users are just everyday folks like you and me. When they slip up, say the wrong thing, or hurt someone's feelings, it's oftentimes unintentional. Not everyone logs online with the intention to hurt someone else. In fact, I would even posit that many cyberbullying attacks only happened because one unassuming person made an unhelpful comment that spiraled out of control and snowballed right into a disaster way bigger than they ever intended.

If you've ever been labeled as "the bad guy" online, then you know just how difficult it is to stop the cycle of hate. In many cases, the barrage of hate that then floods the bully's inbox can be more damaging than the original comment they made. As with all areas of the Internet, it's important to extend forgiveness and compassion—yes, even for the people who annoy you or are all-around jerks. It may seem cliché, but you have to imagine how you would feel before responding to a hurtful message. If it's cruel or unproductive, it may be best to think about the person on the other side of the screen and give it a rewrite.

Despite the prevalence of cyberbullying in our current digital society, I truly feel our new generation is taking more initiative to counteract it than I have ever seen before. Every day, I log online to see dozens of natural digital citizens standing up for others and making their voices known. There will always be the "mean girl" subculture that prevails in certain areas of the Internet; however, I do believe we're entering an era where we do not accept negativity in any spaces and creating a digital culture worth being proud of.

Steps for Managing Cyberbullying

- Be vocal about your firm stance against harmful words and actions.
  - Build an environment around you that stands against cyberbullying and makes your community feel welcome and safe.
- Intentionally curate your online experience to reflect your own morals.
  - Only follow accounts and interact with posts that uplift you and your followers.
  - Avoid online material that harms your self-esteem or promotes hate speech.
- Report bad behaviors.
  - Encourage others to avoid interacting with the negative content and instead block and report the individual enabling it.
- Drown out negativity with positivity.
  - Band together with friends and the rest of your online community to replace the negative comment or post with positive content. Avoid feeding into the aggressor's hostility; instead, counter their aggressions with constructive discussion.
- When standing up for victims, craft a well-written response to the bully, ignore any unproductive responses, and reach out to the victim to provide support.

CHAPTER FIVE

GONE PHISHING

Hello, we've been trying to reach you about your car's extended warranty. Since we've not gotten a response, we're contacting you via this book as a final courtesy warning before we close out your file.

Just kidding. But I wouldn't put it past those automated callers to do something as drastic as trying to glean money from people reading a book. At this point, I'm sure most of us are unsurprised by the measures scammers take to con people. It may be common knowledge—especially for those who grew up on the Internet—to avoid clicking random links, downloading unknown files, or listening to an unidentified person on the phone. However, what happens when the spam message seems real? How deep do these scams go, and what are the scammers hoping to achieve? The answers lie in unpacking what are commonly known as phishing scams, one of the most prominent online dangers out there.

Phishing is the act of sending messages that encourage online users to reveal their personal information, such as credit card numbers or login information. Have you ever received a spam email from an unknown address that reads something along the lines of: "Congratulations! You've won a new car! Click the link to claim your prize: www.thisistotallyarealwebsite.com." Some people may immediately delete an email like this, knowing that if it looks suspicious, it probably is.

Let's say you did click the link. A number of scenarios could happen depending on the sender's goals. Some links lead to a fictitious website designed to trick you into entering your personal information or login credentials, or the link will initiate a download of malware onto your device, enabling the hackers to access your files and control your system from the inside.

Most of us know not to click a random link sent by an unidentified user, and this is doubly true when the phishing message is ridiculous and hard to believe. I recently received a phishing email that went something like this:

> "Greetings,
>
> My name is Mr. Ibrahim Idewu. I work at a bank in Burkina Faso.
>
> I got your contact information from an Internet search. I hope you will not expose or betray the trust I am putting in you for the benefit of both of our families.
>
> I discovered an abandoned fund here in our bank belonging to a dead businessman who lost his life and entire family in a motor accident. I am in need of your help as a foreigner to present you as the next of kin and transfer $19.3 million U.S. dollars into your account. This is 100% risk free.
>
> Please keep this proposal top secret. Also note that you will have 40% of this fund, while 60% will be for me.
>
> This is very urgent. You have five days to give me this information.
> 1. Full name:
> 2. Direct mobile number:
> 3. Home address:
> 4. Job:
> 5. Nationality:
> 6. Gender:
> 7. Age:
>
> Please confirm your interest to receive further information. Please do get back to me on time.
>
> Best regards,
> Mr. Ibrahim Idewu"

The above example is pretty unbelievable, especially when the original email contained countless grammatical errors and odd phrasing, but there are phishing scams that are much more realistic. Some even seem like automated messages from corporations we interact with on a daily basis. Imagine you open your email to this message: "Hello, Your Name. Your USPS package with tracking code GP-3015-JG48 is waiting for you to set delivery preferences: http://bit.ly/36HFjk584." If you have recently placed an order and are awaiting shipment, you may be eager to receive your package and neglect to check that the sender's email address is accurate or if the tracking code matches. Clicking the link would be a simple, yet potentially disastrous, mistake.

Believe it or not, entire companies exist with the sole purpose of designing, developing, and innovating new cyber scams, which means there are constantly new hoaxes to watch out for. In 2021, the FBI's Internet Crime Report depicted the greatest number of complaints and monetary losses to date, with 847,376 complaints filed and $6.9 billion in potential losses. In the same report, the FBI ranks phishing in the top five crimes, on or offline, with most surveyors reporting that they had experienced phishing before.[19] As technology advances, so does its exploitation for nefarious purposes. Criminals are always thinking of new ways to up their game. Understanding the different types of online scams you can encounter, as well as staying up-to-date on any new methods, is Cybersecurity 101. Let's look at some of them in depth.

Hook, Line, and Sinker

There are many different types of phishing (all with equally fun names), but most people regularly encounter two notable types: smishing and vishing. **Smishing** (SMS text messaging + phishing) is a form of phishing where the attacker sends text messages to trick the phone's owner into clicking on a malicious link. For instance, say you received a text from an unknown number that reads, "The IRS is filing a lawsuit against you. An arrest warrant has been issued under your name. For more information about this case file, call the IRS department number at (xxx) xxx-xxxx." Oftentimes, these messages will convey a sense of urgency meant to make you take action without thinking about it first. Even if you do not believe the IRS is filing a lawsuit against you, you may still be confused by the message and will click the link just to see what it is referring to.

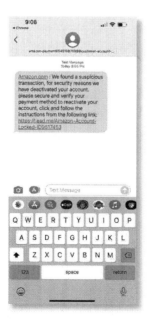

CHAPTER FIVE: GONE PHISHING 63

Likewise, **vishing** (voice + phishing) is a form of phishing that concerns phone calls, an annoyance most of us encounter on a daily basis. Whenever we enroll in online newsletters, create new accounts, or even sign up for webinars, we input our email addresses and phone numbers into a database. This information can then be stolen via the methods we discussed in Chapter 2, as well as bought and sold on the dark web, for the purpose of carrying out phishing scams and placing your number on those call lists.

One of the scariest and oftentimes most effective vishing scams primarily targets older individuals with children or grandchildren. Before enacting the scam, the attacker will access their target's Facebook account or other source of information to learn about their friends and relatives. Then, they will call the target, pretending someone they care about has been arrested to receive money for bail.

Let me paint a picture. Imagine a grandmother, who we'll call Betty, with older teenage grandchildren. Now, imagine she receives an alarming vishing call about her grandson. Betty comes from a time where you could typically trust people on the other end of a phone call, and she certainly has never heard the term "vishing" before. The scenario typically goes something like this:

> SCAMMER: Hello, is this Betty? I am a friend of your grandson's. He was just arrested. I don't know what to do. They're saying they need $2,500 for bail, and I don't have it.

BETTY: Andrew is in jail? What for? He would never do anything illegal!

SCAMMER: Andrew stole something from the gas station, and the owners called the police. He's in a lot of trouble. The only way to get him out is to pay the $2,500.

BETTY: Well, how can I get the $2,500 to you? Can you come to my house, and I'll pay you?

From this point, the scammer will detail all of the different ways Betty can pay them, though they will usually pick an untraceable method such as cash or gift cards. Since the recent popularization of **cryptocurrency**, or a digital currency that does not rely on government regulation, scammers will request payment via crypto since it also does not leave a paper trail. Now, it may seem obvious that you should never give payment information to a stranger over the phone. However, the con artists on the other end of the line are typically very skilled at their jobs. Above all else, they seek to drive you into a panicked state of mind. They do not want you to think rationally about your decision before you make it, meaning even the most level-headed individual can fall victim to the scam if they are convinced a loved one is in danger. Social engineering at its best!

Years ago, my husband and I embarked on our first trip to Las Vegas. We left our two children with my brother-in-law and his family who wanted to take them on a camping trip to Niagara Falls while we were away. My husband and I had just stepped off of the plane in Nevada when we received a panicked phone call from my sister-in-law. She had received a call from an unidentified number, and the person on the other line insisted they had taken my brother-in-law captive and were holding him at gunpoint for ransom money. *No way*, I thought, knowing better. *This is totally an out-of-the-box phishing scam.* Still, my brother-in-law was the one watching my kids. It felt imperative to contact him and verify the truth.

We called and called. He never picked up. My family, all on vacation or visiting another state, was spread out across the country with no way to verify where he or my children were. My husband and I sat in our hotel lobby on the phone with my sister-in-law who, at this point, started to feel panicked, especially when the callers began telling her personal information about my brother-in-law. They knew my brother-in-law's job, appearance, and other details, all of which they used to create a very convincing, very scary story. They didn't know us personally as a family, yet they picked up so much information from social media, the whole scenario felt real.

Fortunately, my brother-in-law did eventually answer the phone. They had been setting up their camper and preparing for their night, so they didn't hear the calls. Crisis averted. Yet, I still think back to that time and feel amazed (and terrified) at how much research the phishers did before calling. In the right scenario, anyone could fall for the scheme. Plus, all of this happened about seven years ago. Phishing scams have only advanced since then, and you better believe cybercriminals have come up with more effective methods over the years. If we are to keep up with them and ensure we do not fall into their traps, we must stay informed, remain cautious, and educate ourselves on all the ways phishing scams can impact us.

Why Catch a Minnow When You Could Nab a Shark?

In addition to targeting everyday people, phishing scams have also evolved to target corporate environments. I recently received an email from a colleague stating how they were on vacation, and they needed me to send $500 in gift cards to a P.O. box in New York as part of a business transaction. The email's heading included the colleague's full name, making the message look convincing. If I didn't know any better, the email would appear to be from the correct sender. Of course, I deleted the email and blocked the address. However, if I did choose to respond, the email would go to the address behind the alias, and they would attempt to confirm the legitimacy of the transaction, potentially using personal information derived from social media. The scammer would also remind me that the purchase would be reimbursed once they returned from vacation, but of course, I would never see that money again.

Cybercrime is a billion-dollar business, and most cybercriminals run it like one. At a macro level, online scams can even impact entire corporations, especially when ransomware is involved. **Ransomware** is a type of malware that is designed to deny access to files on a computer, network, or mobile device for the purpose of collecting a ransom. If the victim does not pay the ransom, they may not be able to retrieve their data, or it could be publicly exposed on the dark web. Clicking a link contained in a phishing email could lead to the execution of ransomware, which then infects the person's system and allows the attacker to gain control over the data on the device or network. The attacker can then encrypt the data they find and offer it back to the victim in exchange for money, a price the victim is willing to pay so the attacker does not publish or delete the data. Basically, the attacker steals the organization's data, locks it behind a door, and sets a price for the key.

With ransomware, there are a couple of factors that could go sideways: A) the corporation could pay the ransom and not actually receive the key, or B) the corporation could pay the ransom and receive the key to their information, yet their data is corrupted and unusable. Despite what you may initially assume, these outcomes are not typical. In actuality, cyber criminals want organizations to know that the data will be returned safe and sound. They actually go to great lengths to make sure their victims feel certain that if they pay the ransom, they'll retrieve their information. I mean, why would the target pay the money if they didn't think the attacker would return their data?

That's why cybercriminals want good reputations. Some of them even have technical support lines to keep up the facade that they are a legitimate organization. Then, if the victim does pay them, they can keep their data and the cyber criminal can retrieve the money. In the past, the whole situation was treated like it never happened, which organizations found appealing as the security of their databases is tied to the trust in their brands. This is why there are now many laws and regulations that require organizations to report data breaches. Gone are the days of sweeping data breaches under the rug. Organizations are being held accountable for protecting customer and employee data.

> **Ransomware**
>
> All your files have been encrypted with ransomware
>
> Send 0.119 bitcoin to ADDRESS. After confirming the payment, all of your files can be decrypted. If you do not make payment in one day, you will lose the ability to decrypt them
> **AND ALL YOUR FILES WILL BE DELETED**
>
> After receiving the payment, we will contact you and give you decryption tools so you can retrieve your files.
>
> email@email.com
>
> **Decrypted Key**

While cyber criminals tend to target top dogs with confidential information they don't want exposed, the average Joe could also fall for a ransomware scam. After clicking on a malicious link, malware is launched and infects the device or network of devices, locking out the users. Then, some kind of message will appear on the screen, detailing instructions on how to pay the ransom and gain access to the device again. Since COVID hit and the workforce transitioned to remote work from home, individuals are now being targeted at a much higher rate since they are more likely to use personal devices to do business. It is important to understand that when you are using your personal device for work or school by accessing both personal data and business/school data on the same device, you are providing a much larger target for cyber criminals.

Throw Them Back in the Water

In January 2022, a Long Island grandmother, who went by the alias of Jean, received a vishing call about her grandson. The scammer, pretending to be her grandson, reported he had been in a car accident and was in jail for driving under the influence. He allegedly needed $8,000 to bail himself out of jail. Jean knew it was a scam, but instead of hanging up, she played the part of the unsuspecting grandmother. She promised the scammer his money as long as his lawyer visited her home to pick up the cash. Then, when the "lawyer" arrived, she handed him an envelope stuffed with paper towels. As he turned from Jean's doorstep to jump back in his car, police emerged from their hiding spots and tackled him to the ground in an epic "gotcha" moment. Jean had called the police to orchestrate the takedown right after hanging up the phone with the scammer.[20]

Jean's story is certainly satisfying. How many of us wish scammers would face justice for targeting the vulnerable people within our communities? However, we can't all be Jean. In fact, the day after Jean's story broke, another story just like it popped up in the news except, this time, the grandmother in the story fell victim to the scam. Days later, similar stories emerged all over the country, and they continue to persist every day. The scammer Jean caught was just one of many. Though she definitely showed him, she put herself at risk in the process. What would have happened if the scammer was armed? What if the police were late to the scene? What if the scammer had the chance to share Jean's home address with a dangerous associate? There are so many factors that could have gone wrong.

The best way to get back at a scammer is, first and foremost, to keep yourself, your information, and your money safe. Even if you know someone is attempting to scam you, do not continue to engage them for laughs. Like I said, scammers are very good at their jobs. The more time you spend with them, the more time they have to convince you to fall into their trap. The best advice I can give you is this: delete the message before opening it, hang up as soon as you receive the call, and throw any suspicious emails into the spam folder. Once you have protected yourself, take action to protect your digital community. Share information about the scam you encountered with others, and let everyone know how to avoid the bait.

Don't Get Hooked: How to Avoid Phishing Scams

- If a phone call or email seems suspicious, it probably is.
  - Trust your instincts, carefully think through the scenario, and contact people you trust to gain their perspectives.
  - If you know for a fact that you're witnessing a phishing scam, do not engage with the scammer in any way.
- Do not click on links from numbers or email addresses you do not know.
- Do not provide any personal or private information over email or phone with unidentified addresses or numbers.

- When encountering a call that seems like a vishing scam, ensure you hang up and check with loved ones before making any impulsive monetary decisions.
  - For example, if someone calls and claims your niece is in danger, hang up and attempt to call her, her parents, or any relative who would know of her whereabouts.
- If an unknown person asks for gift cards, cash, cryptocurrency, or another form of untraceable payment, recognize that if you pay them, you will most likely never be able to retrieve this money.
- Be ready to report phishing scams to the proper authorities. Since many local police departments are not equipped to respond to cyber crimes, the best agency to handle your situation is the FBI. The Internet Crime Complaint Center (IC3) has a direct portal for reporting cyber crimes at https://www.ic3.gov/.

CHAPTER SIX

LOVE AT FIRST SCAM

When I first established my company, Credo Cyber Consulting, I had to fill out all kinds of paperwork to set up a business bank account. On one visit to my bank, my banker asked me several questions related to my business: What is the name of the business? What is your business's objective? What does your business do? I talked for a bit about my professional experience then briefly touched on my specialties as a cybersecurity consultant. Before I could go more in depth, the banker gasped and looked up from the paperwork.

"Really?" she asked. "Oh my goodness, we just had a terrible situation happen the other day." She then told me about how, one week prior, an elderly woman had entered the bank hoping to transfer $10,000 to an individual overseas. Her husband had recently passed away, and in a state of grief, she found a man online who sympathized with her and supported her—all through the Internet. Apparently, she told the banker that she had fallen in love with this person, and since he did not live in the country, she needed to send him money so he could travel to the U.S. and live with her. After hearing this, the banker sensed the woman was falling into a scam and quickly called her associates over to hear her story. For an hour or so, the bank workers attempted to talk her out of the transaction.

Despite the bankers' reasoning, the woman continued to say, "No, no. He really does love me." Seeing the woman wasn't going to budge, the bankers gave up, and ultimately, she went through with the transfer. Five or six days later, the same elderly woman visited the bank again, crying. She asked if she could do anything to retrieve her money since after she sent it, the man stopped responding to her, which is also known as **ghosting**.

Unfortunately, the bank could not do anything to recover the $10,000—the unfortunate outcome in most of these scams.

After the banker finished telling the story, I explained what the elderly woman encountered. She had fallen victim to a **love scam**, a type of scam where someone creates an online persona to trick unsuspecting victims into sending them money or buying them things in hopes they will one day have a real relationship. Love scammers typically target marginalized individuals or people who are suffering from traumatic losses. In the elderly woman's case, she had just lost her husband; the online scammer probably knew this from browsing obituaries online and identified her as an easy target to manipulate. Then, the scammer looked her up on social media, contacted her, and said all of the right things to make her trust them. Again, scammers are good at their jobs. I have no doubt that this scammer took their time talking to the woman and earning her trust. In some cases, a scammer can spend months or even years convincing their victim that they are a real person; this is an extended period of time where they never ask for money just so that, when they do, their target is more likely to cough up the big bucks.

The act of pretending to be someone else online is called **catfishing** (See? I told you they all had fun names). Catfishing is a very simple process to carry out. All someone needs to do is create an email address and a social media account with false information. Then, they can curate a pretend profile with stolen pictures and made-up information. From there, they can contact just about anyone they want and become anyone they want, using the Internet as a shield for their true identity.

The core principles of cybersecurity are referred to as the CIA triad. CIA stands for confidentiality, integrity, and availability. As of 2022, there is no system in place to uphold those characteristics on social media. Sure, celebrities and other well-known users can have their accounts verified on Twitter, but the average person does not undergo an authentication process when creating an account. Creating a profile is simple, and it means you can say anything and be anyone. As a result, the anonymity of the Internet provides the perfect platform for carrying out attacks on unsuspecting people.

It makes sense that dating sites would be the number one place for catfishing to occur. Dating app users are already looking for a relationship and talking to strangers every day; approaching them with a love scam is more natural than messaging someone out of the blue on Facebook. However, such scams can take place in any online environment. No matter who you are or what sites you're visiting, be conscious of who you engage with. Ensure you know who you are talking to, and validate their identity. Follow the age-old adage your mother probably drilled into your head when you were young: don't talk to strangers. Finally, never— and I mean *never*—meet up with anyone you met online when you have no way of validating who they are. Even if you feel they have your best interests at heart, it is a very dangerous game to play when you don't know their true identity.

Stranger Danger

I cannot talk about catfishing without also discussing the dangers that children (and, let's face it, even adults) face when interacting with strangers online. The Internet is the ideal place for creepy people to be, well, creepy. So, navigating online spaces can sometimes feel like a minefield of unpleasant or explicit content. Luckily, there are ways to avoid the mines, though there are some precautions that adults, teenagers, children, and everyone in between need to take.

Ten years ago, kids had a much harder time finding and diving into adult content. Now, they do not have to do much to be exposed to it. In fact, it's an inevitability. If you're on the Internet for long enough, you *will* see something sexual, violent, or otherwise disturbing. This is just as true for children as it is for adults. How could it not be when all of the information in the universe sits in our back pockets? Maybe we don't often think of it that way, but it's the reality of the world we inhabit.

There are so many positives of allowing children access to this database of knowledge. They can research for school, teach themselves life skills, and contribute to their communities. Additionally, this technology (which, by the way, didn't exist as we know it today until a few years ago) also allows parents to better keep track of and communicate with them.

But, of course, there are always downsides. Like it or not, children have a natural curiosity for the world around them, especially for all of the things we tell them they're not quite ready for. The solution to allowing children to explore the wonders of technology—while also limiting what they engage with—

does not lie in stifling that curiosity. Instead, it all comes back to personal responsibility. When parents make the conscious decision to provide their children with Internet access, they have to understand there is always the potential for danger.

Apps, Websites, and Online Spaces to Consider

- Dating apps (including Tinder, Bumble, and Hinge)
  - Dating apps are age restricted; however, a profile can easily be made by anyone of any age. Dating apps are notorious for catfishing scams, but they can also expose users to explicit content. In extreme cases, the apps can even be used for human trafficking crimes.
  - *Tips:*
    - » Never join a dating app unless you are 18 years old or older.
    - » Limit the amount of personal information you share on your profile and via messaging.
    - » Never meet up with strangers in person before verifying their identity.
    - » If you choose to meet up with someone you met online, the buddy system is best. Be sure to let others know where you are, who you are going to meet, and when to expect you to come home.

- Social media platforms
    - Many social media platforms require users to reach a certain age before they can create an account. However, creating a social media account does not require an identity authentication process. Anyone can lie about their age to create a profile.
    - *Tips:*
        » Never sign up for a social media account unless you meet the age requirement.
        » Limit the amount of personal information you post and share. Never share private information.
        » Do not engage with strangers who reach out to you.
- Online chat rooms, such as Omegle, Chatroulette, and Discord
    - There are many chat rooms online that allow users anonymity, which encourages people to say things to others that they would never ordinarily say. Not feeling like their actions will have consequences, users in anonymous chat rooms can be very dangerous to interact with.
    - *Tips:*
        » Always ensure you know everyone you engage with online.
        » Do not interact with anonymous users in chat rooms.

- Mature or voice chat-enabled video games
  - For the most part, video games are harmless fun. However, some contain mature content that can disturb or frighten younger children. Many games also allow voice chat while playing. Since online users are usually selected at random for each game lobby, anyone in the world could speak to the player. As a result, within gamer culture, it's common to shout expletives and crude comments during matches.
  - *Tips:*
    - » Be aware of video game content before playing it or allowing a child to play it. Search online for its rating and if it features any adult material.
    - » Take note of the culture surrounding the game. If it's a competitive, online game, understand that the game's chat will most likely feature mature comments.

Two main groups are at the highest risk of falling victim to love scams: children and the elderly. Though criminals typically target the elderly for monetary purposes, they target children for control over them. I will be referring to these types of criminals as **predators**, or adults who spend time winning over a child's trust to manipulate and take advantage of them. These predators will often **groom** children; in other words, they will spend a significant amount of time making the child rely on them, isolating the child from their friends and family, and using their adult influence as a weapon to receive romantic or sexual attention from the child.

Bark, a parental control software company that helps parents monitor their childrens' online activity, details step-by-step how predators make their victims trust them then use that trust to manipulate and control their actions and thinking.[21]

1. **Targeting**

    If a young adult or adult chooses to forgo locking down their account, they will be exposed to anyone and everyone who chooses to interact with their profile. Billions of people use the most common social media sites. The creepy ones know how to find the most vulnerable users, which are typically young people who feel disenfranchised, marginalized, and voiceless. Predators are counting on these kids to seek out attention on the Internet, so they can then move in and make the individual feel good about themself. As you will see going through this list, predators are masters of psychological warfare. Everything they do is with the intent to make their target feel dependent on them.

2. **Engaging**

    Once the predator has identified their target, they then reach out to the individual. Though they have probably never met this person or at least do not know them very well, they will ask personal questions about their likes and dislikes, relationships with family members, and workplace or school life. They will incessantly DM or text them while flooding their inbox with compliments, praises, and declarations of love. In some instances, the predator will even buy their target gifts—anything to make them feel seen and heard.

3. **Boundary Testing**

    While carrying out step two, the predator may also attempt to gauge the parents' involvement in monitoring the child's devices. In an adult's case, the predator will try to determine how often the adult talks to their friends or family members about their relationship. Either way, the predator will do everything in their power to ensure their conversations remain private.

4. **Isolating**

    During normal, seemingly average conversations, the predator will sneakily work in criticisms of the target's parents or social circle. They will attempt to establish themselves as the most important person in the target's life in an effort to isolate them. When this happens, the target is less likely to approach someone else for help or to discuss their conversations with the predator, thinking that no one will understand their "special relationship" or that someone will try to tear them apart.

5. **Sexualizing**

    Here's where the predator's actions become especially nefarious. After receiving a lot of personal information from the target and establishing a deep bond with them, the predator may begin asking for sexual photos or messages. These requests can seem innocent enough at first but quickly become inappropriate and intrusive. As soon as the predator receives any sexual response from their victim, they can then use that material as blackmail to keep them in the relationship, saying that if the victim does not continue to engage with them, then they will expose their behavior.

    Alternatively, the predator could forgo blackmail and use guilt. They may reestablish how the victim trusts them, loves them, and depends on them. They will further the divide between the target and their family while continually saying that they "cannot live without" the victim.

6. **Controlling**

    The final step is the predator's ultimate goal from the beginning: control over their victim. They will go through great lengths to preserve the emotional manipulation they have worked to concoct. This stage is when it is the hardest for the victim to speak up about their experiences and oftentimes the moment when the predator will prompt the target to meet them alone, whether this is to perpetrate physical harm or, in extreme cases, carry out human trafficking plans.

Most parents can agree that many of our decisions are made on the fly. A lot of the time, you make the best possible choice you can in the moment and hope you did right by your child. Dealing with technological challenges is no different for many families across the globe. Parenthood does not come with an instruction manual. I'm sure many parents don't even know the first thing about all of the websites and games their children explore—they certainly didn't grow up with any of it. Despite the lack of resources, having a plan in place before stumbling upon an Internet issue is for the best. By listening to their kids, acknowledging the different dangers, putting rules in place, and then establishing consequences for breaking those rules, parents can simplify Internet navigation practices not only for their children, but also for themselves.

The Parental Guide to Online Safety

My first piece of advice for parents is to hand your children a contract that details how to properly use their devices, and let them negotiate its contents with you until you reach an agreement that addresses both their concerns and yours. The contract should explain what it means to be a digital citizen, how to have respect for themselves and others online, and what dangers the Internet presents. Like any contract, it should outline the consequences for breaking a clause with a distinct emphasis on how the consequences will not occur because the child has endangered the parent, but because the child has endangered themself. Make sure they know you want them to feel safe, not micromanaged.

I have included a contract within this book as a template for all of my readers who have children, grandchildren, or young people in their lives who they want to support in staying safe online. For my younger readers, consider showing this to your parents as a way to ensure your safety online.

Acceptable Use Policy for Internet Connected Devices

Introduction: The purpose of this document is to provide clear expectations on what we agree to be acceptable use of all Internet Connected devices. This document is needed so that we all understand and agree on the ways we can stay safe when we are connected to the Internet on devices such as cell phones, iPads, computers, and gaming devices. The goal of this document is for us to come to an agreement on what activities are safe to engage in when online and knowing when to get us involved if you feel like something is not safe or suspicious.

Child's Agreements:
1. I agree to conduct myself in a manner with respect for myself and others when I participate in social media platforms online.
2. I will not engage in any form of bullying or negative speak towards others when I am online. This includes social media sites, forums, and gaming platforms.
3. I will agree to allow my parents access to my social media accounts to ensure that my security settings are set properly.
4. I will not accept connections from or follow people on social media that I do not know without permission from my parents.
5. I will give my parents access to my friends and connections on all social media and gaming platforms to ensure I am being safe.
6. I will not give out personal or private information to anyone online for any reason without permission from my parents, this includes location information.

7. If I feel like someone is being inappropriate to me online, I will let my parents know immediately.
8. I agree to have a monitoring application on my phone to monitor the social media platforms I am engaged on for inappropriate content.
9. I agree to give my device and application passwords to my parents.

Parents' Expectations:
1. I will agree to allow you to have a cell phone, iPad, gaming device (select all) that is connected to the Internet if you agree to conduct yourself safely online.
2. I will always be available to you in case you have questions about anything you are experiencing online regardless of how you ended up in that situation.
3. I expect that you will provide me with your device when asked for it.
4. I expect to be provided with the passwords and usernames for all your accounts.
5. I expect that your social media and gaming accounts are always set to private.

Consequences:
If it is discovered that this agreement is violated, consequences may include:
1. Device being taken away for a period of time.
2. Access to device being permanently taken away.
3. Restrictive use of devices.
4. Access to platform being taken away.
5. Other consequences as necessary.

Having access to the Internet through cell phones, iPads, and gaming devices comes with great responsibility. There are many exciting opportunities to be a part of the digital global community, but there are also many dangers when engaging online. It is our primary goal to ensure that you are safely engaging online. We understand that there are going to be situations that may be scary and could be embarrassing, but we all agree that no matter what we will keep communications open without judgment to ensure that you are safe.

Agreed to by: _____
                         Child

Agreed to by: _____
                         Parent

In addition to establishing agreed upon rules, there are also tools that can be used to monitor social media and text messages. Bark, as I mentioned before, is one of the most popular monitoring tools, but many others exist for the same purpose. The software will flag whenever the child sends or receives an inappropriate message or views sensitive material. Plus, you can block certain websites on their devices, and you can even set a time limit for devices, allowing you to control your child's daily screen time.

Though these tools do exist and combat online dangers, the most powerful weapon in your arsenal is proper communication with your child. Maintaining an open dialogue with your child is vital to ensuring they are digesting content that is appropriate for them. The fact is, if your child is online, they will see sexual, violent, or unpleasant material in some way, shape, or form. You can control how they view content at home, but what about when they are at school or a friend's house? If you let them know you're always available for productive conversations about that material, you allow them to approach you with the problem so you both may arrive at a solution together. You have to make them feel safe and comfortable so they know they will not get in trouble for sharing what they have seen or experienced.

In summary, you can draw up a contract, monitor your child's online activities, and hold conversations about online dangers, but in the end, it is up to your child to navigate the Internet safely. All you can do is educate your child, keep an open dialogue with them, and let them know you're present. Then, you have to trust that you've taught them to have enough respect for themselves and others to make good choices and curate a positive digital environment.

How to Avoid Falling for Love Scams

- No matter how good someone is at making you feel like they have your best interests at heart, interacting with them is a very dangerous game to play when you've never met them and don't know what their background is. This is true for people of all ages.
- Set your social media accounts to private mode. If you do not lock down your accounts, you will expose yourself to anyone and everyone who also engages on that site.
- Limit interaction with unverified strangers who connect with you on any platform (not even a professional website like LinkedIn).
  - This is very difficult, especially in an online society that applauds high follower counts. However, engaging with strangers online leaves you vulnerable to many dangers, including monetary scams, love scams, and predators.
- If you're part of a vulnerable group that is primarily targeted by love scams, ensure you remain cautious when connecting with others online. Ask yourself:
  - Have I ever met this person, and have I verified their identity?
  - Do I feel the need to hide my relationship with this person?
  - Could this person have ulterior motives by talking to me?
  - Has this person used any of the aforementioned tactics to earn my trust?

- If you're a parent, there are many steps you can take to foster a positive online environment for your child.
    - Use parental monitoring software to limit the types of websites and content your child can access.
    - Access your Wi-Fi router's settings to filter the types of websites you can visit when connected to your home's network.
    - Most importantly, foster a healthy, honest relationship with your child; encourage open communication about what they're engaging with online.
        » Establish Internet safety rules with a written contract to facilitate essential conversations about online dangers and how to stay safe.

CHAPTER SEVEN

FAKE NEWS OR NEWSWORTHY?

When I was growing up, the primary way to receive information was from the library, the five o'clock news, or other materials like newspapers, books, or magazines. Research was far more intensive and purpose-driven. If you wanted to learn more about a topic, you better have been ready to do some work. You had to jump in your car and drive to the library. Then, you had to look up the book or resource, and God forbid the book was already checked out! Writing papers for school involved doing some leg work, and you *know* we went through a heck of a lot of Wite-Out.

The way we engaged with news was also completely different. News reporters had street cred. They spent their whole careers building up reputations by going out into the trenches and doing some hard-hitting investigative journalism. They couldn't just buy a microphone and camera to begin broadcasting; no, the Carl Bernsteins and Barbara Walters of the world had to put in the work first. Then, the nonfiction texts, books, and articles they put out for public consumption had to be screened for accuracy and fact-checked. I'm not saying it wasn't impossible to hear something wrong on the news, but the idea of "fake news" certainly wasn't the dilemma that it is today.

Today, a 14 year old with an iPhone or their own website can reach just as many people, if not more, than an educated, experienced journalist. There's something to be said here about how search engines filter and share content via algorithms, which we will definitely touch on later, but there's also a point to be made about human curiosity. When we see something surprising, intriguing, or relevant to our interests, we tend to want more information. Then, we want to share this information with others to form those oh-so-vital social bonds or improve our own understanding of a topic. But what happens when that information isn't as accurate as we thought it was? What happens when it's flat-out wrong?

Curiosity is part of human nature. How we feed that curiosity is when we start to run into issues. If you don't know something or, in many cases, want to prove someone wrong in an argument, it's likely you turn to Google for answers. Now, Google is a revolutionary technology that has completely transformed how we find information... but therein lies the problem. Google is not a subjective platform—it's designed to provide users with a tailored content experience. The results you see when you search with keywords may not be 100% true or based in fact. Instead, the results reflect what Google thinks you want to see. This aspect of Google, paired with website algorithms and targeted advertising, allows people to essentially form information bubbles around themselves. This creates a very dangerous society where people are unwilling to learn new facts, hesitant to accept opinions outside of their own, and complicit in the spread of misinformation and disinformation, or as we have so affectionately dubbed it in the last couple of years, fake news.

Misinformation vs. Disinformation

The difference between misinformation and disinformation comes down to intent. **Misinformation** is when someone inadvertently shares false information. Maybe someone retweets a surprising news article without reading it only to find out that all of the information in the article is untrue. Alternatively, someone could make a bunch of posts about a subject without knowing all of the facts, fully believing that they know the truth. Misinformation is primarily a concern on social media feeds as people tend to share content without first checking if it comes from a credible source.

Disinformation, on the other hand, is when someone *deliberately* shares incorrect information with the goal of changing someone's mind or impacting how a group views a topic. There's currently an extremely concerning issue involving nation states that create what appear to be real news outlets to distribute myths and disinformation. The news outlet may claim to be Canadian but really be part of the Russian GRU. The people behind the accounts even create social media profiles for the outlets, and by purchasing bot followers, they can appear credible and well established on a surface level. Once an account has a blue verification check next to its name and 100,000 followers, many people will believe what it has to say without question.

The Future of Fake News

Fake profiles may also include Photoshopped images that make the account look more trustworthy. However, Photoshop changes are sometimes easy to recognize, especially if the image is poorly edited. Plus, with the rise of memes and the accessibility of image-editing software, many people understand that they shouldn't trust every picture they see on the Internet. But what happens when it isn't a picture that has been edited, but a video? A new, scary, and mind-blowing trend is sweeping the Internet called **deep fakes**, or falsified videos where a person in one video is replaced with another person's face, body, and voice. If you search "deep fake" on YouTube, you will see dozens of videos of famous people, such as Morgan Freeman, Tom Cruise, and even former U.S. president Barack Obama, saying outlandish things they would never say in real life. These videos look real, but they aren't. Instead, they are made with artificial intelligence (AI) that reviews hours and hours of video footage to learn a person's mannerisms, the inflections in their voice, their facial expressions, and more.

Deep fakes are the future of fake news. The videos are already concerning enough when most people do not fully know how to create them, but the technology will only become more advanced and accessible, meaning deep fake technology has the potential to become just as significant as Photoshop.

Imagine a world where you can't trust a video of someone. Deep fakes could be used to damage someone's reputation, making it seem like they said or did something harmful. They could be used as cyberbullying tactics, given the person creating the video takes footage from the victim's TikTok or Instagram content. They could even be used as robotic technology to give human-looking robots more realistic facial expressions, body movements, and gestures. Currently, we don't know how widespread deep fake technology will be, but we do know this: deep fake technology in its current state is just the beginning.

As if it wasn't hard enough to navigate the current Internet minefield of fake news, we also face another form of disinformation known as clickbait. **Clickbait** is content that is created to entice people to click on an over-the-top, exaggerated, and untrue news article. You know you've struck clickbait when you see a headline like "Meryl Streep Caught Hiding Pregnancy? Shocking!" or "This Pill Melts Fat in Your Sleep." Most of the time, clickbait links do not even lead to real articles. Instead, they could lead to an ad for a product, or worst-case scenario, they could bring you to a page hiding malicious software, which could lead to issues with malware, ransomware, or information theft. Either way, someone on the other side of the screen typically makes money from the number of clicks a piece of clickbait generates. The more enticing the headline, the more money the creator makes, so of course making up an outrageous lie would generate more clicks than a real news story!

Clickbait is commonly found on real news sites. Sometimes, the links are even mixed with real news articles, so telling the difference between real and fake takes a keen eye. Take a look at Figure 7.1 and Figure 7.2 for example. The articles are divided into squares that each lead to a different page. Notice how some feature a CNN logo in the top right corner while others do not.

Figure 7.1

Figure 7.2

The real article links look very similar to the clickbait ones. In fact, I wouldn't be able to tell the difference between the two if not for the logo in the top right corner of the thumbnail images. Plus, the clickbait headlines are noticeably different from the more serious and formal articles from the real news site. In the example, the clickbait appears in the same spot as the real news articles, but they can show up anywhere on the page, including on the sides of the article in their own sections. Many times, the clickbait will also have a headline over it that reads "Paid Content" or "Sponsored by (Website Name)."

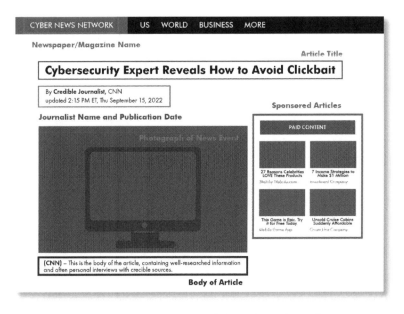

I've also noticed a recent trend that I heavily endorse: some news sites have added an advertiser's disclosure to the top of their articles. When you hover your mouse over the link, a paragraph pops up, detailing how they receive money from third parties to display clickbait and other articles as advertisements.

This disclosure immediately signals to readers that content from outside sources appears on the page, so even if they trust the news site, they need to be cautious when engaging with all of the links present. Just because it shows up on a reputable site does not mean the link itself is safe. If you see one of these advertiser's disclosures out there in the wild, know that clickbait could be lurking just around the corner—or, in this case, just down the page.

Before falling into the clickbait trap, always hover over the link first to verify its authenticity. When you do, a box will appear under your mouse that shows where the link leads. Ask yourself: have you heard of the website before? Does the web address look suspicious? Even if the article looks enticing and you just *have* to know what terrible thing Jennifer Aniston said to David Schimmer while filming *Friends*, it's best to avoid any unfamiliar websites. The best way to protect yourself against clickbait (and, really, all online dangers) is to think before you click. Be intentional with what you engage with on the Internet. Even if it's difficult and takes extra time, it's necessary. In my experience, though, avoiding clickbait isn't the hardest part; avoiding the algorithms is.

Don't Get Caught in the Algorithm Loop

The rise of algorithm-based Internet experiences, meaning platforms that now cater content to individuals based on what they engage with online, is scary. Personally, I'm the type of person who always wants to hear both sides of an argument, debate, or story. I like to see what information is out there, hear all of the angles, and make informed decisions based on my own morals and principles. Algorithms are making this process increasingly more difficult.

When you open a browser to search for a topic or scroll through your social media feed, algorithms only show results based on the millions of data points that have been gathered about you from your previous online engagements. This means you'll probably only see content that is aligned with what you have engaged with in the past, and finding opposing opinions without independently seeking them out is a challenge. When all you see is information that backs up your thoughts, you become trapped in an **algorithm loop**. It's hard to see the other side of the story; you may not even know there *is* another side. This is a dangerous problem that feeds into an "I'm right and you're wrong" mindset, which can lead to divisiveness.

Algorithms used in websites, search engines, and social media platforms are designed and implemented for curating personalized experiences online. In 2000, Google's web crawler and PageRank algorithm revolutionized information retrieval. The notion behind these mathematical computations included on and off-page factors. If there were external links pointing to a page, then Google gave that landing page a higher rank. The thought process behind these computations was "if people are talking about you, you must be important."

In 2007, Facebook introduced the like button as its first experimentation with algorithms. In addition to the like button, Facebook used the X-out feature to tailor the user's news feed based on likes and content that was previously closed. This was the beginning of customized user experiences and how we arrived in the current state of getting caught in the algorithm loop. You know you are in one when you watch one cute puppy video, and for the next several days, you are bombarded with cute puppy videos. Awesome!

But what if you see a post joking about how the Earth is flat? Let's say you become curious and start Googling the history of Flat-Earthers. You dig a little deeper, and before you know it, Twitter starts recommending that you follow some accounts who post theories about the Earth being flat. Maybe you follow them as a joke, but pretty soon, your whole feed is nothing but flat Earth propaganda. When everyone in your bubble only talks about how Flat Earth is real, you could definitely start to believe the theory, especially when your search engine prioritizes showing you blog posts and articles about it. After all, *everyone* is saying it's true, so it must be, right?

Taking Back Control

Anyone who has surfed the web knows how easy it is to fall into an algorithm loop. Thankfully, there are steps and precautions you can take to ensure you don't accidentally brainwash yourself. The first step, as always, starts with personal responsibility. Escaping the hold algorithms have over our online activities takes some work, especially with how quickly algorithms evolve. You have to take responsibility for the information you learn and actively seek out facts. All of this should be done with the goal of hearing both sides of the story and then drawing a conclusion for yourself.

You can help end the rise of fake news. When you see a new article on your Facebook feed, you shouldn't just read the headline, accept it as the truth, scroll past, and then repeat that information whenever the topic comes up in real life conversation. Instead, you should read the headline, determine whether the news site and journalist are credible sources, and read the entire article if you want to learn more. The fun doesn't stop there! You should also search for articles with opposing viewpoints as well as articles with neutral positions. Then, you can determine what you think about the topic.

The search for articles should extend beyond a quick Google search. As we've discussed, Google's responses are tailored to your likes and dislikes, but it also features sponsored links. If I search for Ford trucks, then Dodge trucks will definitely also appear on the top search results page as a sponsored link. Dodge, as well as many other companies, pays good money to have its website links appear at the top of Google's search results pages. What do you think happens when most people don't visit the second or third pages of Google results when conducting their research? That's right: a ton of potential information gets pushed down on the results page, cast aside, and ignored in favor of sponsored results.

In a perfect world, learning how to identify and avoid fake news would be part of every school curriculum, especially when finding credible, research-backed sources is as easy as finding blog posts or opinion pieces. You just have to know where to look. When you are scrolling through search results on Google or your preferred search engine, looking for information, note that some websites are more trustworthy than others. When you visit a website while gathering information, be sure that the URL is accurate and not misspelled. Oftentimes, bad actors will spoof a website by making a new site that appears very similar to a credible one. For example, the American Red Cross' URL is https://www.redcross.org, but a bad actor could create a website called https://www.red-cross.org and lead unsuspecting users to a malicious site. Also, if the site represents an organization, most of the time it will have a .org at the end of the URL. Educational institutions and government websites tend to have .edu and .gov at the end of their URLs respectively. However, remember that just because a site has .org, .edu, or .gov at the end of the URL doesn't mean it is actually a credible site.

When you click on an article, pause to see if the article says "opinion" at the top or not, and know that blog posts will most often be opinion pieces unless they contain links to credible sources. Google Scholar, a search engine that only brings up scholarly sources, also acts as a perfect companion to your Google search and an alternative to Wikipedia. Anyone can visit a Wikipedia page and edit its contents, so even though Wikipedia pages are heavily monitored by the website's editors, any academic organization worth its weight in salt will not allow a Wikipedia page as a cited source. However, the links and sources on Wikipedia pages can lead to many valuable resources for you to explore, so consider starting your research there and ending with a pile of scholarly information.

All of this research requires more work than a simple Google search. This aspect of digital citizenship is absolutely one of the most time-consuming, yet it is also one of the most necessary. Being intentional about what you consume online is essential to ensuring you do not spread false information or aid in some large-scale disinformation scheme. It is also important on a personal level. By leveraging your critical thinking skills to consider your own thoughts, morals, and opinions, you help yourself become a more thoughtful and well-informed individual who isn't quick to jump on a bandwagon. Then, the culture you build around yourself spreads to others within your circle of influence, creating a more thoughtful and educated online society.

Tricks for Avoiding Fake News

- Rule number one: do not use Facebook and Twitter as your news outlets.
  - Seven out of ten U.S. Twitter users say they use the site for news, especially to follow breaking news as it unfolds, though only 7% of these users claim to have a "great deal" of trust in the news' accuracy on Twitter.[22]
  - As of 2020, about a third of U.S. adults use Facebook for news, which was even higher than the number of people who use Twitter.[23]
- Read an article before sharing it. Don't simply read the headline and accept it as fact.
  - While you're at it, verify that the news site and journalist are credible sources, not an opinion blog or WordPress site.
- Recognize what is and what is not a credible source.
  - Credible sources can be found using Google Scholar or another academic database.
  - Wikipedia is not a reliable source. However, you can use it as a jumping off point for your research.
- Understand the difference between clickbait and real articles. Avoid falling into the clickbait trap no matter how appealing the headline may seem.
- Analyze how social media algorithms are tailored to you. If you start to notice that you're seeing the same kinds of content over and over, consider branching out to see alternative thoughts, opinions, and content.

- Before forming an opinion on a person, situation, or event, put in the time to hear from every side of the story.
  » Read articles from the side you think you agree with. Read articles from the side that opposes your initial thoughts. Then, read articles that seem to only present facts.
  » Form your own opinion once you have all of the details instead of letting headlines manipulate your mindset.
- Because we're being bombarded with millions of data points every single day from all these different sources, it's important to slow down, take a minute, and evaluate the content you're consuming. Does it always seem to fit into one narrative or viewpoint?
  » Being intentional in engaging with online content is not easy. However, it is necessary work, especially with so much misinformation and disinformation in circulation.

CHAPTER EIGHT

STRIKING THE DIGITAL BALANCE

Many people reading this right now either purchased or were gifted this book because they spend way too much time on the Internet, or as kids refer to it, they are "chronically online." I don't mean to criticize anyone for their Internet habits. I mean, can I get a raise of hands from people who are tired of hearing about how phones are rotting our brains? Technology now impacts every part of our lives, and the fact of the matter is we have to live with these devices. In many cases, turning the phone or computer off could mean missing a vital message from a loved one, getting lost on the way to an important meeting, or not being able to call 911 in the case of an emergency. We depend on technology for so much, and let's face it: it's not going anywhere. So, how do we go about using our devices for all of these necessary aspects of our lives without also spending too much time plugged in? The answer lies in finding the right digital balance for your own needs.

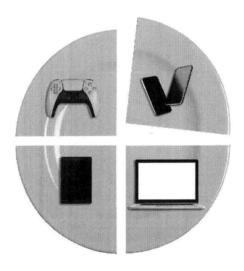

I think of digital balance in the same way I think about a balanced diet. Just like how you wouldn't eat an entire chocolate cake every day for dinner, you shouldn't spend excessive amounts of time on the Internet. Diets are different for everyone; they largely depend on current eating habits, physical activity, and other characteristics that differ from person to person. Finding a digital balance is no different. The right balance for you will depend on three questions: what, when, and how much?

- What do you consume?
    - The digital content you consume can greatly impact your perspectives, opinions, and even mental health. Taking note of the content you read, watch, and engage with is the first step to determining what content is beneficial and what content can be cut out.
- When do you consume it?
    - It's easy to pick up your phone as a way to procrastinate, but doing so can waste a lot of time that could be better spent on productive or more fulfilling activities. By considering when you spend the most time online, you can take note of if your habits are interfering with other aspects of your life.
- How much is too much?
    - If you find that the time you spend on your devices interferes with your productivity, physically or mentally impacts you, or negatively affects your relationships with others, it may be time to reevaluate your browsing habits.

What, When, and How Much?

At one point in my life, I, as a wife and mother, pursued my master's degree with a full-time job while volunteering for seven different industry organizations at once. People would always ask me how I found the time and how I could be everywhere all at once. Juggling all of my responsibilities was difficult for sure, but I was able to "find the time" by changing my habits and cutting recreational content consumption out of my daily routine. For two years, I rarely watched television, and I purposefully did not sit on social media for extensive periods of time. Sure, I sat in front of a screen a lot while I was at work and researching for my thesis, but my technology usage was deliberate. As a treat to myself, I would schedule time to surf my favorite social media sites and news outlets. But, I would control the amount of time I spent online by turning off all notifications except for work and school related content. That's the key difference between healthy digital intake and digital addiction: deliberate consumption.

Digital addiction, also known as Internet addiction disorder, is the continual act of mindless content consumption when online for long durations of time. People with digital addictions have a dependence on technology and find it difficult to stop engaging with it despite the harmful effects they experience. An example could be sitting on YouTube for seven hours a day or playing video games for several days straight. Even if a person begins to feel negative consequences, such as deteriorating vision or back problems from poor posture, their habits do not change. More severe consequences have even begun to be studied further in depth with smartphone usage being linked to increased levels of anxiety and depression,[24] trouble sleeping,[25] and increased risk of car accidents.[26] As of 2022, digital addiction is not acknowledged as a mental disorder in the DSM-5, but the topic of whether or not to include it has been largely debated. Internet gaming disorder, however, has been mentioned as a condition that warrants further research.[27]

A study from Pew Research claims that 31% of U.S. adults say they go online "almost constantly," a percentage that has increased from 21% in 2015.[28] When we interact with technology, particularly with social media and video games, our brains release dopamine,[29] a neurotransmitter that carries messages between neurons, nerves, and cells in the body. Dopamine "rewards" us for our actions and plays a part in forming habits, which is why it's so hard to put the phone down and disengage.[30] Since we can easily access this release by reaching in our pockets and pulling out our phones, we train ourselves to crave the rush it provides, creating an addiction. This is why children and teenagers, whose brains are not yet fully developed, are at the highest risk for digital addiction, as they naturally have higher dopamine responses.[31]

What makes all of this even more complicated is that many apps and websites are *designed* to be addictive. Site creators want users to form habits around their products so the users will revisit the site again and again. Consider the layout of most social media apps. The content never ends; you can keep scrolling forever and never reach the bottom, or in the case of YouTube and other streaming services, the next video automatically plays. This is so you never have a moment to pause and think,

"Maybe I should do something else."[32] Some apps even hide the time at the top of your screen, wanting you to lose track of time, or they will pride themselves on only featuring **microcontent**, or very short snippets of content to keep the dopamine flowing. Think of Twitter where the word limit is a measly 280 characters or TikTok where most videos are under a minute long. Basically, every aspect of an app's design, from the way the content is portrayed to the color choices, are intentionally chosen to keep you interacting with the content.

Now more than ever, it's important to be intentional about the content you consume. The amount of time you spend online should not only be deliberate but mindful. The tactile feel of a phone in your hand can feel like a safety blanket at times or even a way to break an awkward silence. Waiting rooms, grocery store lines, the DMV—the list of places where anyone would want a distraction goes on and on. However, the more you can separate yourself from your technology, the better.

Weaning Yourself Off

Prior to the COVID-19 pandemic, people had moments in the day where they could not look at a screen. If you were in a meeting or a class, it was downright rude or sometimes prohibited to glance at your phone or laptop, and face-to-face interaction was encouraged. Now, with the popularization of remote working and learning environments, technology is often required for that interaction, meaning the amount of time people spend on their devices has drastically increased. Before, there was a level of disconnect between work life, school life, and home life, whereas now, professional and personal have somewhat blended together.

Because of this, there's no standardized amount of time everyone should spend online per day; the right digital balance is different for everyone and depends on an individual's lifestyle. A YouTuber will spend way more time on the Internet than a soccer coach, so how they monitor their online time will differ. Someone may earn their living on their home computer or attend online classes that are required for their diploma, and this time spent online cannot be simply cut out of a routine. That being said, regulating your screen time starts with taking note of how much you cannot cut out versus how much you can. Then, you can set realistic goals for yourself to lessen your technology usage.

As someone who has worked remotely for more than a decade, the most helpful way for me to limit my screen time is setting boundaries. I personally turn off all social media notifications on my phone when I work. This way, I do not become distracted and waste time looking at yet another screen, which enables me to use the time I would have otherwise wasted on my phone for other tasks throughout the day. By taking note of how much time I spend online, I can then monitor how much time I could put towards activities I'm passionate about. How many of us wish we had more than 24 hours in a day to pursue our hobbies or accomplish our goals? Most of the time, we can reclaim some hours by rethinking our online habits.

Using the calendar app on my phone, I also set alarms for certain times in the day. I have one alarm that rings every day at the same time to let me know it's time to hit the gym. Then, during my workout, I try not to look at my phone or any other screens; it's blocked out as time for myself. If I didn't do this, it would be very simple to fall into mindless content consumption. Twenty minutes here and there throughout the day can easily add up to several extra hours spent on a device. Sometimes, you can intentionally replace that time with a productive task. Instead of browsing social media, maybe you could finally find the time to go for a walk, spend time with your kids, play your favorite sport, or read a book you've been looking forward to. I guarantee you'll find you can get a lot more done in a day than you originally thought.

No matter what you do, remember this: the key to limiting your screen time is not judging yourself, no matter how much time you spend on technology. As I always say: no ego, amigo. Everyone needs to grow, and everyone is always learning. You can't beat yourself up about it; you just have to sit down, examine your current screen time, make actionable goals, and stick to them. Fortunately, there are many different programs and apps that exist to help people track and manage their screen time. I personally use a program called Toggl Track on my computer to see how much time I spend on my laptop as a whole as well as certain work projects. As with anything, do your research to see what software will work best for your needs. There are all kinds of options available.

In addition to computer software, most smartphones come with a built-in monitoring feature. iPhones and Androids have similar features that provide you with daily and weekly summaries of how much time you spend on your phone and which apps you use the most. From these features, you can also choose apps to close automatically during certain times of the day, as well as see how many times you unlock your phone and how many notifications you receive in a day. The only difference between iPhone and Android is where to find the feature. Under the "Settings" app, the option is located under "Screen Time" for iPhone users and "Digital Wellbeing" for Android users.

Recently, I began busting my husband's chops for being on his phone too much. So, I asked him to open his Digital Wellbeing feature to check his screen time. Seeing he spent four hours a day on his phone, I opened my phone to show him my screen time, expecting it to be much lower. Low and behold, our screen times were basically identical. This caused me to take a minute for self reflection—I never thought our screen times would match! How could this be? I needed to sit down and examine what apps and websites I was spending my time on. I discovered that I spent much of my time scrolling through Instagram and LinkedIn, while the rest of my time was used for YouTube and my daily meditation app. After thinking about it for a while, I ended up running a little experiment to see just how much I could reduce my screen time. I turned off social media notifications and set more calendar alarms. Most notably, I made a conscious effort to put down my phone when I picked it up without thinking. Next thing you know, my screen time dropped from four hours per day to 46 minutes per day. Now, I'm more aware of how much time I spend on my phone doing non-productive scrolling, and I am conscious of how people around me engage with their technology as well.

In summary, we need to take back control and hold ourselves accountable for our time. As individuals, it's necessary to face the hard truth: our relationships with ourselves and others can be impacted by our technology usage. In pop culture, digital addiction is not treated with the same sense of urgency as other addictions. I personally believe this is because a majority of people suffer from some form of digital addiction, whether they acknowledge the role technology plays in their lives or not. For many people, being more present in their own lives would change a lot about their perspectives and give them more peace. The only way to regain this control is through deliberate content consumption, not mindless scrolling or passive dilly-dallying. Ask yourself: does the content I consume make me feel good, and how is that feeling serving me? If your life isn't better for it, maybe it's time to break a habit or two.

How to Hit the Empower Button

- To determine if your digital habits are impacting your life, ask yourself:
    - What do I consume?
    - When do I consume it?
    - How much is too much?
- Download screen time monitoring software or use apps to track how much time you spend online. Many options exist, so do your research to find the one that best suits your needs.
    - By determining how much time you spend on your devices, you can then create an actionable, reasonable game plan for reducing your screen time that corresponds with your personal lifestyle.
- Set clear boundaries for yourself.
    - This could include creating an alarm that tells you to engage in a nontechnical activity like working out or taking a walk, or you may set your phone to limit access to certain apps after you spend a specified amount of time on them.
- If you're struggling to hit your digital balance goals, buddy up. Find a friend or family member who will keep you accountable for your technology usage. Teamwork makes the dream work!

- For some, an effective way to begin the withdrawal process is to quit cold turkey. Delete the apps you spend the most time on for a set amount of time. Then, you can slowly begin reintroducing them into your life with new goals and limitations in place.
  - An example of this would be someone deleting Instagram from their phone, not looking at the app for two weeks, then setting boundaries that say they can only look at it for 30 minutes per day.
- Above all else, be honest with yourself. No matter what amount of time you spend online, do not cast judgment or feel guilty. These platforms are built to suck you in and make you spend as much time on them as possible. You just have to be stronger than the marketing psychology.

CONCLUSION

The human race is on the cusp of some very significant technological changes. The way we engage with the entire world is shifting right before our eyes. On top of the natural progression we experienced pre-2020, the COVID-19 pandemic acted as a catalyst for technology, making all of us more reliant on it for shopping, communication, and healthcare, among other things. The best innovations are created out of necessity, and boy, did we ever need new software, devices, and solutions during the pandemic!

We already have so much machine learning and artificial intelligence in our lives, including household names like Siri, Alexa, and Cortana. I can open my phone with facial recognition technology. We couldn't even do that ten years ago! I certainly never dreamed my phone would be able to sense my face from any angle—or with a mask covering my mouth and nose! In addition to facial recognition software, there are also innovations with fingerprint recognition, touch screen swiping, and a host of other advancements. As we find ourselves adapting to remote work and school from home, we are also moving toward more secure, passwordless environments, which is crazy exciting for those of us in the cybersecurity community and also for the everyday digital citizen. And this is only one example of how society is progressing. There are so many more advancements happening behind the scenes.

The progress we're making is both exciting and scary at the same time. In an ideal world, every technologist would only develop digital solutions for the betterment of society as a whole. We'd see spectacular gains in science, medicine, and automation, helping experts in all industries move the world forward. New technologies would assist people with disabilities, decrease the amount of work required in labor-intensive professions, and help find the cure to some of the most fatal diseases on the planet. While all of these advancements and so many more are certainly being researched and explored by countless organizations and specialists, we unfortunately do not live in a perfect world. When a ground-breaking technology emerges, someone somewhere will find a way to use it for wrongdoing.

As digital citizens, it is now our responsibility to ensure we ethically use these new technologies and stay aware of how others can use them to harm us or our communities. There will always be positive and negative actions we can take. One positive action is maintaining an upstander mentality and strong values when engaging with these new devices and programs. Then, by staying informed, hopefully the positive actions can start to overpower the negative ones and contribute to a safer Internet community.

After reading this book, I hope you feel empowered to take control of how you engage with others online. I hope you understand the value of cyber hygiene as well as how to implement it in your own life—not just to help yourself but to become a productive, caring member of the digital world. The tools within this book are actionable. They're not too technical. They're not too difficult. You can do this, one tip at a time.

At the end of the day, holding ourselves accountable for our online habits and then working to improve our behavior will only lead to self-betterment. By understanding the role technology plays in our lives, we can begin taking intentional steps toward being digital citizens who curate a positive, safer online society. An informed digital citizen can easily recognize clickbait because they conduct their own research and apply critical thinking skills to think before they click. They witness cyberbullying, on large or small scales, and actively oppose the mob mentality by being an upstander who speaks out for victims. They know how their digital footprint impacts their real lives, and most importantly, they inform their loved ones of the dangers, which could potentially save them from all kinds of threats and scams.

The human aspect of cybersecurity has always been what called me to the profession. When I originally attended college to pursue a criminal justice degree, I quickly learned how much passion I had for physical security. I loved the idea of building secure environments for people. Then, when I went back for my master's degree, I began seeing truths about technology I couldn't unsee, truths that continue to threaten the global landscape. In the past, countries only fought wars with physical armies and weapons. Now—and I don't mean to sound like an alarmist—digital wars are being waged against average citizens participating in online spaces. These wars take the shape of misinformation and disinformation, data monetization, information theft, and so much more. The most disturbing part is most people don't even know how these dangers impact their lives. Upon learning all of this for myself, I knew I had to be part of the solution.

In the first company I ever worked for, my boss gave me two pieces of advice: 1) either you're part of the solution, or you're part of the problem and 2) read the freaking manual (affectionately abbreviated to RTFM). I knew the lack of cybersecurity was a major problem, so I had to make a decision. Should I put blinders on and pretend like everything is okay? After all, I'm only one person. What difference can I make? Or, should I proactively take a stand and do my best to make digital environments better for everyone?

The choice seemed obvious and is a huge part of why I wrote this book in the first place. Part of my mission is to give all generations their power back when engaging online. In fact, I originally started my company, Credo Cyber Consulting, as a resource for K-12 environments and small to medium-sized businesses that don't have the same resources as large companies. I want everyone, no matter if they're part of an organization or just an individual, to be able to say with confidence, "I have control over my data. I understand the dangers I can encounter on the Internet. I'm prepared not only to face them but continue learning about them so I can teach others how to face them as well."

If I had all of the money in the world, I'd spend the rest of my life evangelizing for educating young people and adults alike in the importance of proper cyber hygiene. I'd be a leading advocate for anyone who doesn't know where to start with computers or smartphones, let alone how to be safe while fiddling with them. On top of it all, I'd encourage all school systems to implement curriculum changes from kindergarten to 12th grade. Our education system needs to reflect how technology impacts every aspect of our lives. Math class could integrate lessons about coding languages and algorithms, social studies could discuss disinformation and misinformation, English could teach proper communication when on social media—the possibilities are endless. By incorporating small bits and pieces into every subject, our children's knowledge of technology would be just as impactful as the technology itself. I truly, truly believe that this is the only way we will start mitigating some of the threats within our current global online landscape.

The large-scale change I'm talking about cannot happen overnight. It will take time for every digital citizen to take responsibility for their online activity. The information and tips listed in this book are not simply suggestions; they're the beginning of a movement that starts with you, the reader. The first step is awareness. The next is action. Begin by integrating some small changes to your digital routine. Then, little by little, introduce some more. Before long, you'll have built a digital fortress around yourself that is more positive, more productive, safer, kinder, and above all else, a testament to all the good we can do if we just come together and take action. From there, your little digital community will cause a chain reaction that will affect our digital landscape on a global scale. Your ability to create positive change in the online world is what truly makes you a global citizen.

Daily Cyber Hygiene Checklist

This checklist features easy, actionable tasks you can do right now to begin protecting yourself online. Proper cyber hygiene is a continual process, but these are steps you can take today to start your cybersafety journey.

- ☑ Complete daily social media housekeeping by being mindful of the content you're consuming.
- ☑ Set up private and personal social media accounts.
- ☑ Read every website's privacy statement. Don't just click "I Agree."
- ☑ Avoid personality quizzes, as they are typically tools for websites to collect your personal information.
- ☑ Use a password vault service to store your accounts' usernames, passwords, and security questions.
- ☑ Restart your phone every day to close out any apps you may have had open, reducing the risk of them being compromised.
- ☑ Use a VPN when connecting to public Wi-Fi networks.
- ☑ Be aware of how your location settings are set up across the apps on your phone, and change any location settings that seem intrusive or unnecessary.
- ☑ Enable multi-factor authentication whenever possible using an authentication app as the third factor rather than SMS text messaging.
- ☑ When visiting any website in your browser, look for the lock icon next to the URL, which indicates that the website is secure.
- ☑ Use the wellness or screen time applications on your phone to monitor how much time you spend online.
- ☑ Think before you click.

Simple Tech Tools and Best Practices

Security Tools:

- Use a password vault service, such as LastPass, 1Password, or Dashlane, to store your accounts' usernames, passwords, and security questions.
- Download antivirus and anti-malware software, such as McAfee and BitDefender, to protect your devices against malicious attacks.
- Subscribe to a VPN service, such as Private Internet Access, ExpressVPN, NordVPN, or Surfshark, to hide your IP address and protect your privacy online.
- Set up Microsoft BitLocker on Windows to prevent unauthorized users from accessing your hard drive's data even if the hard drive is removed from the computer.
- Implement multi-factor authentication on all of your accounts, including social media, banking, and other applications you use to protect against brute-force attacks.
- Use an authentication app, such as Google Authenticator or Microsoft Authenticator.

Best Practices:

- Enter Google Chrome's "Incognito Mode" or Firefox's "Private Browsing" mode when accessing your browser to navigate the Internet privately.

- Download a search engine or browser like DuckDuckGo that does not collect or share personal information.
- If you're a parent, use parental monitoring software, such as Bark, to limit the types of websites and online content your child can access.
- Get a digital certificate that validates your identity on your email to ensure your name cannot be effectively used for phishing scams.

ACKNOWLEDGMENTS

I have to start by thanking my incredible family. To my sons, Tyler and Dylan, and my husband, Gary, thank you for your endless support throughout this process. When I told my husband I was going to write a book, he looked at me and said, "Awesome! Go for it!" Thank you for never once doubting me. From listening to me go on about the latest chapter I was working on to helping me select the right colors for my cover art, the three of you have always made me feel like this project was just as important to you as it was to me. I love you all and could not have completed this book without you by my side.

To my niece, Mikayla. Thank you for your input on the draft manuscript. Your ideas were awesome, and I am so grateful you took an interest in helping me with this project. I cannot wait to see where your cybersecurity career will take you. Just know that I will be front and center to witness it all.

If you are lucky enough in life, you get to grow up with siblings. I was blessed with two. My brother and sister are the only two people on this planet who share a lifetime of experience with me. Janet, when we were kids, you always treated me as your equal. You have been my best friend for as long as I can remember, even through our ups and downs. Thank you for supporting me in everything I do. Anthony, nobody can make me laugh the way that you do. Thank you for always bringing humor into every situation to remind me not to take life so seriously. I love you both.

Siblings are amazing, but to have in-laws that are as close as it comes to siblings also rocks! Steve and Erica, we have been rolling together for almost 30 years. You have both been a constant throughout my life. Thank you for believing in me even when I didn't always believe in myself. To my niece and nephew, Adrianna and Andrew, thank you for taking such an interest in my book and for being my test subjects for the cover art. Your feedback was invaluable to me. I love you guys.

Thank you to the incredible team at Brightray Publishing. Jamie, without you I don't think I could have completed this project. You have been with me every step of the way, and I am grateful for our time together and your insights. You are wise beyond your years, and you have been a joy to work with. Nicholas, your patience with me during the art selection process was greatly appreciated. I didn't realize there were so many different fonts to choose from. Thank you for guiding me through it and helping me to land on the perfect one.

To Jeffrey Slotnick, CPP, PSP, thank you for being a mentor and a friend. That fateful day when you sat down at the "ladies lunch table" during the ASIS Leadership Conference in D.C. and introduced yourself to the group, I knew that you would be an ally and someone I wanted to get to know better. You have always been a champion for women in the security industry and that does not go unnoticed. Thank you for your kindness, for your sage advice and wisdom, and for introducing me to your incredible wife, Nancy. You are truly the dream team, and I am grateful to know you both.

To my mentor and colleague, Jeroen Kouwenhoven, thank you for the hours that you spent listening to me, helping me navigate my dreams, and always providing me with loving feedback that was honest. You knew how to deliver the message even when I may not have always liked what I was going to hear. The lessons you have taught me will stick with me forever.

To my team at Moms in Security Global Outreach (MISGO), especially Janet, Elisa, and Min, thank you for supporting me in all that I do. Working with you in the fight against human trafficking and child exploitation has been part of the motivation to complete this project. Bringing awareness and education about the dangers of the Internet to young people and their families is a big part of helping them protect themselves from being victimized. I hope this project will aid in the prevention of these terrible crimes.

To my friend and colleague, Heidi, thank you for always being a sounding board both professionally and privately. I cherish our friendship and the growth we have had together over the past five years. Keep sending me those podcast suggestions!

To my security family, of whom there are too many to list, thank you all for always pushing me to be the best version of myself. Being a part of this community has been one of my biggest joys. From my time early on in my career as a technician to starting my own firm, there were so many of you that guided me, supported me, and celebrated the milestones with me. Thank you for being my sounding board, my confidants, and my friends.

Introduction

1. "The State of Mobile in 2022: How to Succeed in a Mobile-First World As Consumers Spend 3.8 Trillion Hours on Mobile Devices," Data.ai, January 12, 2022, https://www.data.ai/en/insights/market-data/state-of-mobile-2022/
2. Bertha Coombs, "Loneliness is on the rise and younger workers and social media users feel it most, Cigna survey finds," updated January 23, 2020, https://www.cnbc.com/2020/01/23/loneliness-is-rising-younger-workers-and-social-media-users-feel-it-most.html

Chapter Two

3. Matthew D. Lieberman, *Social: Why Our Brains are Wired to Connect* (New York: OUP Oxford, 2015).
4. Mariana Gabi et al., "No relative expansion of the number of prefrontal neurons in primate and human evolution," *Proceedings of the National Academy of Sciences* 113, no. 34 (2016): 9617-9622, https://doi.org/10.1073/pnas.1610178113
5. Louis Cozolino, *The Neuroscience of Human Relationships: Attachment and the Developing Social Brain* (New York: W. W. Norton & Company, 2014).

6. Debra Umberson and Jennifer Montez, "Social Relationships and Health: A Flashpoint for Health Policy," *Journal of Health and Social Behavior* 1, no. 51 (2010): S54, https://www.ncbi.nlm.nih.gov/pmc/articles/PMC3150158/
7. "Money makes the cyber-crime world go round— Verizon Business 2020 Data Breach Investigations Report," Verizon News Center, May 5, 2020, https://www.verizon.com/about/news/verizon-2020-data-breach-investigations-report
8. Jennifer Kastner, "In-Depth: Up to $1K offered on dark web for patient medical records," May 11, 2021, https://www.10news.com/news/in-depth/in-depth-up-to-1k-offered-on-dark-web-for-patient-medical-records

Chapter Three

9. Sam Meredith, "Facebook-Cambridge Analytica: A timeline of the data hijacking scandal," April 10, 2018, https://www.cnbc.com/2018/04/10/facebook-cambridge-analytica-a-timeline-of-the-data-hijacking-scandal.html
10. "How do I permanently delete my Facebook account?" Meta Help Center, accessed July 11, 2022, https://www.facebook.com/help/224562897555674
11. "What is the Privacy Policy and what does it cover?" Meta Privacy Center, last modified January 4, 2022, https://www.facebook.com/privacy/policy/

12. "How to deactivate your account," Twitter Help Center, accessed July 11, 2022, https://help.twitter.com/en/managing-your-account/how-to-deactivate-twitter-account

13. "Privacy Policy," Snap Inc., June 29, 2022, https://www.snap.com/en-US/privacy/privacy-policy

14. Melissa DeCesare, "Moms and Media 2021," May 6, 2021, https://www.edisonresearch.com/moms-and-media-2021/

15. "Parents 'oversharing' family photos online, but lack basic privacy know-how," Nominet, September 5, 2016, https://www.nominet.uk/parents-oversharing-family-photos-online-lack-basic-privacy-know/

16. Rosie Hopegood, "The perils of 'sharenting': the parents who share too much," October 11, 2020, https://www.aljazeera.com/features/2020/10/11/facing-the-music-the-parents-who-share-too-much

Chapter Four

17. Justin W. Patchin, "Bullying During the COVID-19 Pandemic," September 29, 2021, https://cyberbullying.org/bullying-during-the-covid-19-pandemic

18. "Toxicity During Coronavirus Report—L1ght," L1ght, accessed July 11, 2022, https://l1ght.com/Toxicity_during_coronavirus_Report-L1ght.pdf

Chapter Five

19. "Internet Crime Report 2021," Federal Bureau of Investigation, accessed July 11, 2022, https://www.ic3.gov/Media/PDF/AnnualReport/2021_IC3Report.pdf

20. "'Bored' Long Island Grandma Teams Up With Police to Bust Scammer Targeting Elderly," 4 NBC New York, January 21, 2022, https://www.nbcnewyork.com/news/good-news/bored-long-island-grandma-teams-up-with-police-to-bust-scammer-targeting-elderly/3508480/

Chapter Six

21. "What is Grooming? Signs to Look for With Sexual Predators," The Bark Blog, January 3, 2020, https://www.bark.us/blog/grooming-signs-sexual-predators/

Chapter Seven

22. Amy Mitchell, Elisa Shearer, and Galen Stocking, "News on Twitter: Consumed by Most Users and Trusted by Many," November 15, 2021, https://www.pewresearch.org/journalism/2021/11/15/news-on-twitter-consumed-by-most-users-and-trusted-by-many/

23. John Gramlich, "10 facts about Americans and Facebook," June 1, 2021, https://www.pewresearch.org/fact-tank/2021/06/01/facts-about-americans-and-facebook/

Chapter Eight

24. Kadir Demirci, Mehmet Akonui, and Abdullah Akpinar, "Relationship of smartphone use severity with sleep quality, depression, and anxiety in university students," *Journal of Behavioral Addictions* 2, no. 4 (2015): 85-92, https://akjournals.com/view/journals/2006/4/2/article-p85.xml

25. Sakari Lemola et al., "Adolescents' Electronic Media Use at Night, Sleep Disturbance, and Depressive Symptoms in the Smartphone Age," *Journal of Youth and Adolescence* 44 (2015): 405-418, https://link.springer.com/article/10.1007/s10964-014-0176-x

26. Erich Muehlegger and Daniel Shoag, "Cell Phones and Motor Vehicle Fatalities," *Procedia Engineering* 78 (2014): 173-177, https://scholar.harvard.edu/files/shoag/files/cell_phones_and_motor_vehicle_fatalities.pdf

27. Ranna Parekh, "Internet Gaming," June 2018, https://www.psychiatry.org/patients-families/internet-gaming

28. Andrew Perrin and Sara Atske, "About three-in-ten U.S. adults say they are 'almost constantly' online," March 26, 2021, https://www.pewresearch.org/fact-tank/2021/03/26/about-three-in-ten-u-s-adults-say-they-are-almost-constantly-online/

29. M. J. Koepp et al., "Evidence for striatal dopamine release during a video game," *Nature* 6682, no. 393 (1998): 266-268, https://pubmed.ncbi.nlm.nih.gov/9607763/

30. Simon Parkin, "Has dopamine got us hooked on tech?" March 4, 2018, https://www.theguardian.com/technology/2018/mar/04/has-dopamine-got-us-hooked-on-tech-facebook-apps-addiction

31. Birgitta Dresp-Langley, "Children's Health in the Digital Age," *International Journal of Environmental Research and Public Health* 17, no. 9 (2020): 3240, doi:10.3390/ijerph17093240

32. Catherine Price, "Trapped—the secret ways social media is built to be addictive (and what you can do to fight back)," October 29, 2018, https://www.sciencefocus.com/future-technology/trapped-the-secret-ways-social-media-is-built-to-be-addictive-and-what-you-can-do-to-fight-back/

GLOSSARY

Algorithm loop: A situation where you find yourself stuck in a loop of reading only the information that backs up your thoughts without being exposed to the other side of the story. When you log on, search for a topic, or scroll through your social media feed, algorithms only show results based on your current interests, which makes falling into an algorithm loop much easier.

Brute-force attacking: A hacking technique where cybercriminals buy an individual's PII on the dark web, giving them access to their accounts across all websites.

Bystander: A person who stands by and takes no action when witnessing cyberbullying or another form of online negativity.

Catfishing: The act of pretending to be someone else online.

Clickbait: Any text, link, or content that is created to entice people to click on an over-the-top, exaggerated, and untrue article or an unrelated website.

Cookies: Small text files that are dropped into your browser when you engage on a website. They are also convenient tools for website creators to track and save information about the time users spend on the site.

Cryptocurrency: A digital currency that does not rely on government regulation.

Cyberbullying: A situation in which one person feels threatened, targeted, or harassed by another person when engaging in online spaces. Cyberbullying can happen to or be perpetuated by any digital citizen via any online medium, including texting, apps, and video games.

Data monetization: A process in which a company sells your information to advertisers who then use it to create targeted ads.

Dark web: A collection of hidden online content only accessible via a special web browser called Tor. Criminals use the dark web for a wide variety of tasks, including the illegal selling of information.

Digital addiction (also known as Internet addiction disorder): The continual act of mindless content consumption when online for long durations of time.

Digital citizen: Anyone in the global digital community who actively participates in the use of technology, engages with others via cyberspace, and consumes information and data in every aspect of their life.

Digital DNA: The permanent trail of information an individual leaves behind when browsing the Internet.

Disinformation: Incorrect information that is deliberately shared with the goal of changing someone's mind or impacting how a group views a topic.

Ghosting: The act of ending all online communication with someone.

Grooming: A term used to describe the coercive methods used by predators who use their adult influence as a weapon to receive romantic or sexual attention from a child.

Hate speech: Any public statement that encourages discrimination against a group of people based on their race, ethnicity, sexuality, gender, or other identities.

Identity theft: A crime where someone wrongfully obtains and uses PII to commit fraud or other crimes. An identity thief could use stolen information to illegally apply for a credit card, file taxes, or receive medical services.

Love scam: A type of scam where someone creates an online persona to trick unsuspecting victims into sending them money or buying them things in hopes the scammer and victim will one day have a real relationship.

Malware: A type of software that intentionally harms your device. It can be encountered in various places across the Internet but most notably via phishing scams and sketchy websites.

Metadata: The information that is stored about an online user, including their IP address, physical location, computer hardware, router type, personal information, and other identifying characteristics.

Microcontent: A very short snippet of content on online or social media apps.

Misinformation: Any incorrect information people accidentally share online because they believe the information is true.

Multi-factor authentication (also known as dual-factor authentication): An authentication method in which a user is only granted access to an account if they can supply two or more pieces of evidence verifying their identity.

Password vault: A service which offers highly encrypted, online safes that hold people's account information and generate strong passwords for them.

Personal information: Any information that is not unique to you but is still a part of your personal life, such as the name of your school or workplace, your eye color, or your favorite sport.

Phishing: The act of sending messages that encourage online users to reveal their personal information, such as credit card numbers or login information. Most phishing links lead to a website designed to retrieve a user's information, or the link will download malware onto their device, enabling the hackers to access their files.

Predator: An adult who spends time winning over a minor's trust to manipulate and take advantage of them.

Private Information (also known as Personally Identifiable Information, or PII): Any information unique to you, such as your social security number or credit card information.

Ransomware: A type of malware that threatens to publish a victim's data and information to the world unless they pay a ransom.

Sharenting: A term used when a parent publishes content surrounding their children on social media or other online platforms.

Smishing: A form of phishing where the attacker sends text messages to obtain the phone owner's personal information.

Upstander: A person who stands up for others and takes action to defend and support cyberbullying victims.

Vishing: A form of phishing where the attacker tries to obtain personal information through phone calls.

ABOUT THE AUTHOR

Antoinette King, PSP, DPPS, founder of Credo Cyber Consulting, LLC, has more than two decades of experience in the security industry. Drawing on her experience, Antoinette founded Credo Cyber Consulting in 2020 with the goal of providing her clients a holistic perspective on security, bridging the gap between the physical and cybersecurity domains with a focus on data privacy and protection.

Antoinette has always seen security as her vocational calling. She has been passionate about educating people on the importance of staying safe online for many years, especially in the corporate setting. However, after being asked to present at a digital learning day for fourth and fifth grade students, she recognized that young people are far more in tune and engaged with what is happening online than most adults realize—or understand, for that matter. It was this experience that was the driving force behind writing this book. Her goal is to be a resource for students, teachers, parents, aunts, uncles, and grandparents by creating educational content that is easy to understand and actionable.

Antoinette's certifications include Board Certified Physical Security Professional (PSP), certified Data Privacy Protection Specialist (DPPS), and SIA Security Industry Cybersecurity Certification (SICC). Her degrees include A.S. in Criminal Justice, B.S. in Managing Security Systems, and M.S. in Cybersecurity Policy and Risk Management. Antoinette has published and presented on various security and cybersecurity topics in several capacities, including articles, white papers, keynotes, webinars, panel discussions, and corporate talks.

Made in the USA
Middletown, DE
25 April 2023

29360277R00091